PROCEEDINGS
OF THE
TWENTY-THIRD
ANNUAL MEETING
8–9 APRIL, 1987

NEW DOSIMETRY AT HIROSHIMA AND NAGASAKI AND ITS IMPLICATIONS FOR RISK ESTIMATES

As Presented at the Auditorium of the National Academy of Sciences Building

Issued December 31, 1988

National Council on Radiation Protection and Measurements
7910 WOODMONT AVENUE / Bethesda, MD 20814

The material included in these proceedings has not been reviewed and represents the views of the authors and not necessarily those of the NCRP.

Preface

The theme of the 23rd Annual Meeting of the NCRP was limited to the topic of the "New Dosimetry at Hiroshima and Nagasaki and Its Implications for Risk Estimates." This subject provides the basis for the experience of the Japanese survivors of the atomic bombs dropped in 1945 which make a major contribution to our knowledge of the risk of late effects (cancer) in human beings from exposure to ionizing radiation. Critical to these risk estimates is the dosimetry of those exposed and revisions in our appreciation of this dosimetry have been in progress for a number of years but have recently been completed as reported in this volume.

The meeting began by addressing the new dosimetry itself, some perspectives and an overview of the physics involved, then some special aspects of the dosimetry followed by the details of the new dosimetry system (DS86) and its application to the survivor information at the Radiation Effects Research Foundation. Discussions on the second day focussed on the impact of the new dosimetry with respect to several biological endpoints and especially on risk estimates for cancer. The program committee selected distinguished speakers on each of these topics, all of whom were contributors to the development of the new dosimetry and its application. A special feature of the program was the contributions made by Japanese scientists to the work and to the program.

These proceedings also include the text of the Eleventh Taylor Lecture "How to be Quantitative about Radiation Risk Estimates" presented by Seymour Jablon. The speaker was introduced by Gilbert W. Beebe.

The meeting closed with brief reports from the work of three NCRP committees on Radon Measurements, on Quality Assurance and Measurements in Diagnostic Radiology, and on Genetic Risk of Internal Emitters.

The Council is indebted to the authors who participated in the program and provided the material published here and to the James Picker Foundation for past support of the Taylor Lecture Series and NCRP Publications. The Council also acknowledges its debt to the members of the Program Committee for the 1987 Annual Meeting.

R.J. Michael Fry, Chairman
Oak Ridge National Laboratories
Oak Ridge, Tennessee

Edward L. Alpen
Lawrence Berkeley Laboratory
Biology and Medicine Division
Berkeley, California

Gilbert W. Beebe
National Cancer Institute, NIH
Clinical Epidemiology Branch
Bethesda, Maryland

Victor P. Bond
Brookhaven National Laboratory
Medical Department
Upton, New York

William C. Roesch
Battelle Pacific Northwest Laboratories
Richland, Washington

William J. Schull
University of Texas
Houston, Texas

I would also like to express my personal thanks to these persons and to Laura Martin of the NCRP Staff for compiling this volume.

Bethesda, Maryland
June 1, 1988

Warren K. Sinclair
President, NCRP

Contents

TWENTY-THIRD ANNUAL MEETING OF THE NATIONAL COUNCIL ON RADIATION PROTECTION AND MEASUREMENTS

The New Dosimetry at Hiroshima and Nagasaki and Its Implications for Risk Estimates

Opening Remarks

Warren K. Sinclair
President, NCRP

I am very pleased to welcome here today all the members of the Council and its committees, the representatives of collaborating organizations and all others interested in the work of the NCRP, to this 23rd Annual Meeting.

May I draw your attention to the pamphlet labeled NCRP news which is available at the registration desk. It notes the six documents that NCRP published last year and indicates two others that will shortly be available. The Annual Report is also available to those who wish it and it contains detailed information on the NCRP program, and especially about additional NCRP reports soon to be published. Among these are two documents of special importance, one on "Recommendations on Limits for Exposure to Ionizing Radiation" which will be the first of its kind from the NCRP in sixteen years. The other is on the "Exposure of the Population in the United States to Ionizing Radiation" and is an assessment of all sources of radiation to which the population is exposed.

The scientific program today is of special importance to all those interested in radiation protection because it concerns the new dosimetry at Hiroshima and Nagasaki and its implications for risk estimates.

From 1975 onwards questions were raised, initially by the NCRP, about the validity of the T65 dosimetry used as a basis for describing the exposure of the survivors of the atomic bombs dropped at Hiroshima and Nagasaki. These led to provisional revised estimates of the doses by Loewe and Mendelson at Lawrence Livermore National Laboratory and by Kerr and Pace at Oak Ridge National Laboratory in 1980–81. These revisions were discussed at a Symposium at the Radiation Research Society's meeting in Minneapolis in May 1981. Subsequently, a broadly based program involving examination of all aspects of the dosimetry was undertaken under the auspices of the Department of Energy with support from the Department of Defense. J. Thiessen played a key role in its sponsorship. R. Christy of the California Institute of Technology led a U.S. team of investigators entrusted with this responsibility. A corresponding Japanese team led by E. Tajima directed the studies in Japan. Senior Scientific Committees at the National Academy of Sciences in the U.S.A. chaired by F. Seitz and in Japan by E. Tajima reviewed the work of the investigative teams. The investigations, which range from new detailed calculations to new measurements of the gamma rays by thermoluminescence dosimetry and involve modifications in allowances for shielding by structures and of organs, are detailed in a volume presently being prepared for publication by the Radiation Effects Research Foundation (RERF) in Hiroshima, Japan. In addition, scientists at the Foundation are engaged in applying the new dosimetry to the data base of survivor information at RERF and thereby assessing the impact of the dosimetry on risk estimation in the Japanese survivors.

Dr. Michael Fry (of Oak Ridge National Laboratory) and his program committee have arranged a scientific program here today which will address all aspects of this subject. I am very pleased to introduce Dr. Fry to you at this time to provide some additional information on the program.

Scientific Session

The New Dosimetry: Perspectives and Overview

Eizo Tajima
Chairman

Early Work Carried Out by Japanese Scientists

Tatsuji Hamada
Japan Radioisotope Association
Tokyo, Japan

1. Introduction

On the 8th of August in 1945, two days after the bombing at Hiroshima, Dr. Y. Nishina, nuclear physicist famous for the well-known Klein-Nishina formula for Compton scattering of photons, went to Hiroshima at the request of the Imperial Headquarters. There, he found enormous damage and concluded it was due to an atomic bomb, as he had suspected prior to his visit. This conclusion was supported by fogging of developed x-ray films stored at the Hiroshima Red Cross Hospital and by induced radioactivity detected in various materials collected at Hiroshima.

Two days later, Prof. B. Arakatsu of Kyoto University and Prof. T. Asada of Osaka University also went there.

On September 14, the Science Research Council of Japan organized a large scale research group to study the effects of the atomic bombs in Hiroshima and Nagasaki from the viewpoints of physical and biological sciences. More than 250 scientists and physicians took part in the study which was done over a few years following the bombing. The investigations were carried out under the extremely difficult circumstances in the early post-war days in Japan. Instruments for radiation measurements were, of course, not so well developed as they are at present. Even the project of study cannot be said to have been well-organized. The scale of damage in both cities was much beyond all imagination. Moreover, typhoons hit Hiroshima twice and killed 11 members of the group from Kyoto University.

The research group prepared the reports [1] consisting of 1,580 pages, and there were 38 papers even in only the physics and technology section. The reports, published in 1953 in two volumes, can be regarded as one of the important literatures in the history of science, though most of the 38 papers are unsatisfactory for dosimetric purposes.

In this article a brief review will be made of the results of these and subsequent studies carried out by Japanese scientists until 1965, especially those which contributed significantly to the present dose reassessment.

2. Determination of Hypocenter and Height of Detonation

M. Kimura and E. Tajima [2] of the Institute of Physical and Chemical Research (IPCR) determined the hypocenters and the heights of detonation in Hiroshima and Nagasaki using shadows of objects cast on surfaces of wood, concrete and granite by the intense thermal radiation from the fireball of the explosion. Figure 1 illustrates the result for Hiroshima. The heights of detonation were estimated to be 577 ± 20 m for Hiroshima and 490 ± 25 m for Nagasaki.

3. Thermal Radiation Measurements

Kimura and Tajima [3] collected, at a distance of 920 m from the epicenter in Hiroshima, a wooden board of Japanese cypress, the surface of which was charred by the thermal radiation of the explosion. They distinguished two layers, a completely carbonized layer and a brown, incompletely carbonized layer, and measured their thickness. Then, they made an experiment using an electric furnace and a carbon arc furnace to reproduce these conditions of charring, varying the temperature of the heat source and the time of exposure to the thermal radiation from the heat source.

From the result of the simulation experiment, the thermal energy fluence at the location was estimated to be about 41 cal/cm^2, which corresponds to a total radiant energy from the bomb of 4.6×10^{12} cal assuming the bomb and the wood to be the black bodies and neglecting attenuation of the thermal radiation in the atmosphere.

T. Watanabe *et al.* [4] of the University of Tokyo studied thermal effects on rocks and ceramic products at Hiroshima and Nagasaki from the petrological and geological point of view. They determined the range of distance from the hypocenter where exfoliation of granite or melting of roof tiles were observed. The data for Nagasaki is presented in Figure 2. Exfoliation of granite is considered to be due to the phase

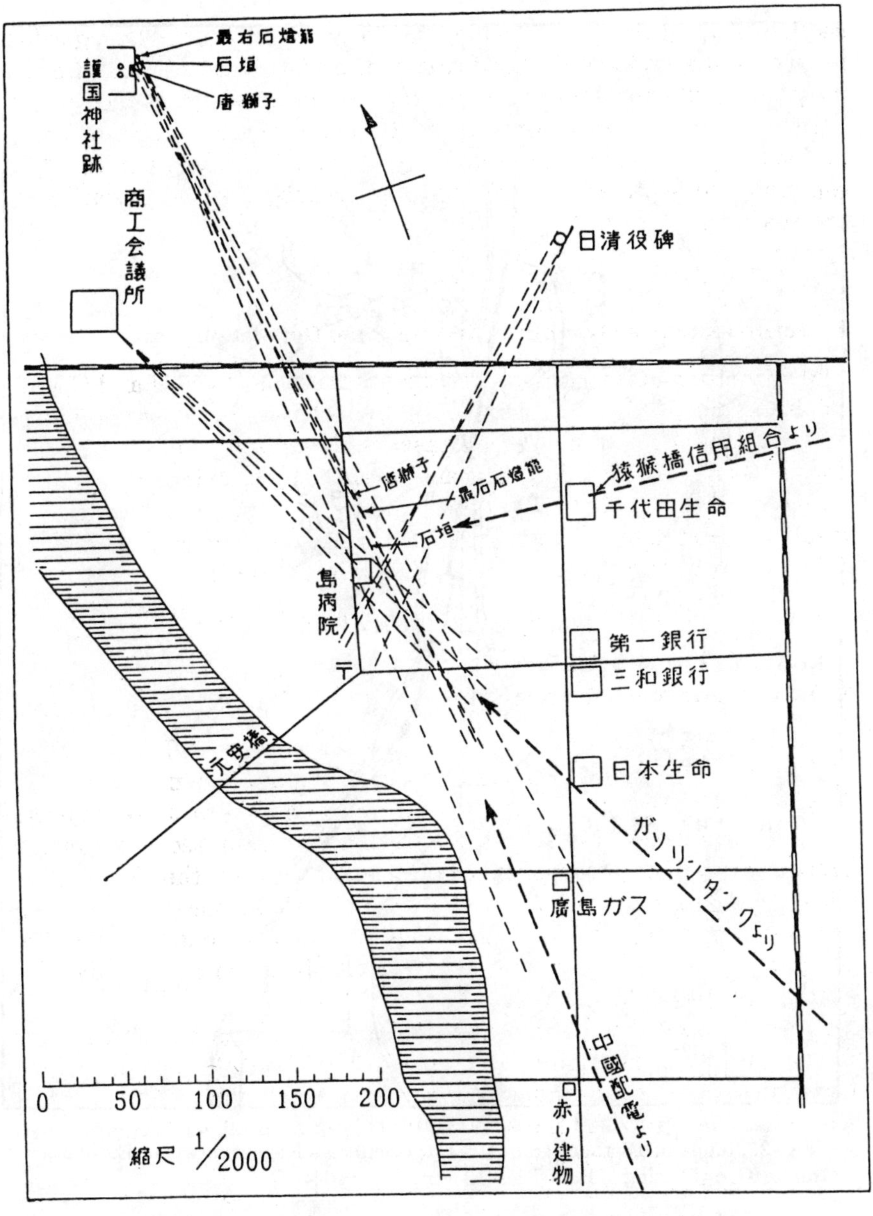

Fig. 1. Determination of the Hypocenter at Hiroshima [2].

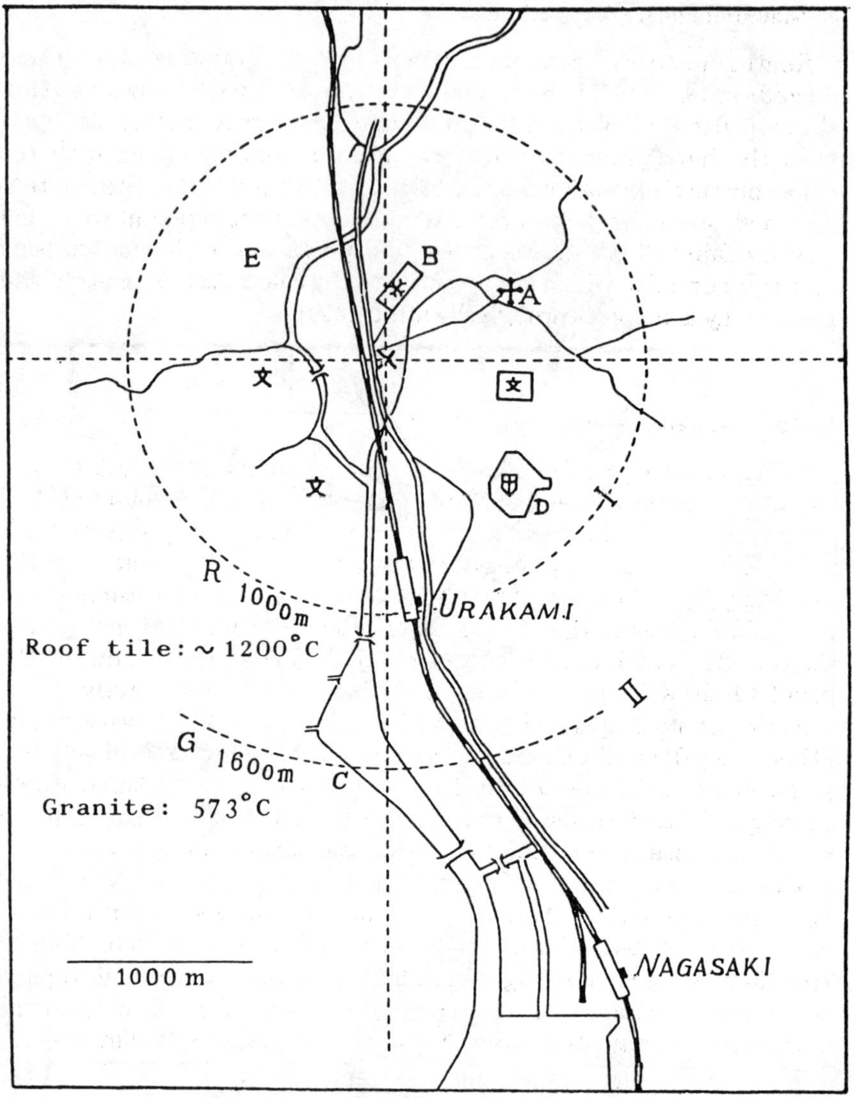

Fig. 2. Range of Distance from the Hypocenter at Nagasaki, where Exfoliation of Granite (G) or Melting of Roof Tiles (R) were observed [4].

transition of quartz at 573°C, while the melting point of roof tiles was about 1,200°C.

The data for roof tiles were used later by Tajima [5] for the estimation of the bomb yield in Hiroshima relative to that in Nagasaki.

4. Meteorological Observations

Routine meteorological observations [6,7] at ground level were conducted on the day of the bombing at the Weather Stations in Hiroshima, 3.6 km SSW from the hypocenter, and in Nagasaki, 35 km E from the hypocenter. Recently, these data were combined with the tropospheric observation data obtained at Koenji Weather Station that had made aerological observations on a routine basis at that time, and used by Tajima [5] to estimate the atmospheric lapse rates for temperature and humidity which are essential to the calculation of neutron and gamma ray transport through the atmosphere.

5. Residual Radioactivity

F. Yamasaki [8] of IPCR measured the activity of a dust sample collected from the gutter of a house located at Takasu, about 4 km W of the hypocenter in Hiroshima, where major fallout was observed. He found that the activity decayed according to $t^{-1.31}$, where t is the number of days following the detonation. From this, he estimated the maximum exposure rate 1 h following the bombing, exposure during the period from 2 h to 24 h after the bombing, and exposure during the next 24 h to be about 20 mR/h, 220 mR and 40 mR, respectively.

K. Shinohara *et al.* [9] of Kyushu University, and T. Fujiwara and H. Takeyama [10] of Hiroshima University of Education made direct measurements of exposure rate on the ground with Lauritsen electroscopes at Nagasaki and Hiroshima, respectively. Their results are, unfortunately, expressed as multiples of the natural background.

Miyazaki *et al.* [11] of IPCR made measurements with a Neher type high pressure argon chamber used for the routine observation of cosmic-ray intensity. Their results obtained in Jan. to Feb. 1946 at Hiroshima are shown in Figure 3. The figures indicate the ionization in terms of "J" which corresponds to 1.5 μrad/h. They also found an ionization of as high as 670 J at Nishiyama in Nagasaki in Jan. 1964.

6. Induced Radioactivity

A number of data are available on the induced radioactivity of materials at the sites. Among them, those for ^{32}P in sulfur are important because they provided the basis for estimating fast neutron fluence. F. Yamasaki and A. Sugimoto [12] of IPCR collected sulfur samples from pole insulators at Hiroshima and measured their activi-

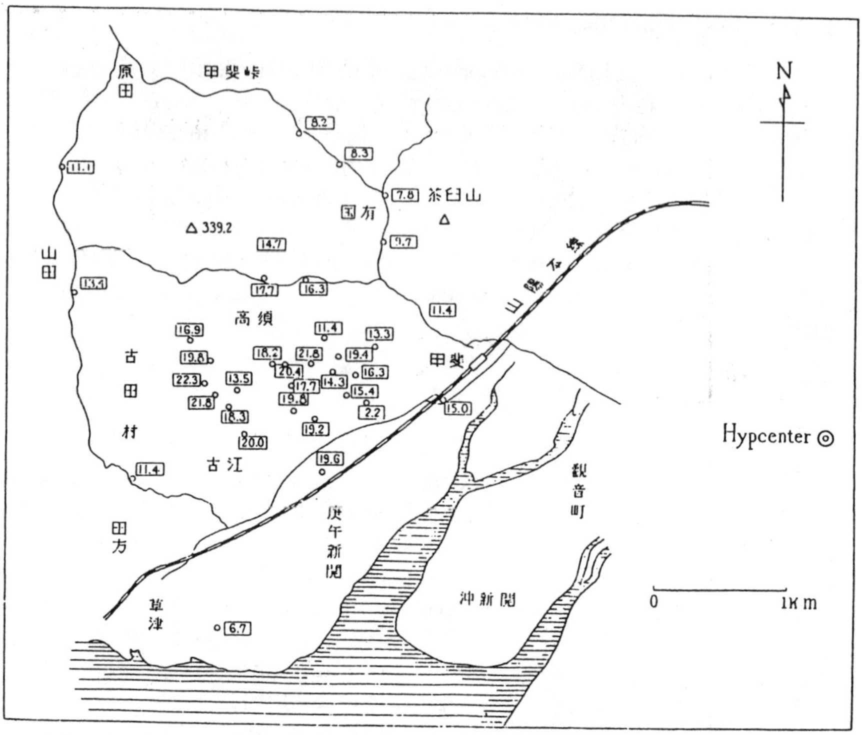

Fig. 3. Readings of a Neher Type Ion Chamber Obtained at Takasu, about 4 km W of the Hypocenter at Hiroshima in Jan. to Feb. 1946 [11]. The Figures Indicate the Ionization in "J" Unit.

ties using a Lauritsen electroscope calibrated with a standard source of uranium. The results are presented in Figure 4 as a function of the distance from the hypocenter.

The results were reexamined by Yamasaki [13] in 1958 and the fast neutron fluence above 3 MeV was estimated to be $2.8 \times 10^{10}cm^{-2}$ at the hypocenter without any correction for neutron attenuation in ceramic material of the insulator. Tajima and N. Oda [14] experimentally determined the attenuation factor using fast neutrons from accelerators and obtained the factors 0.52 and 0.77 depending on the direction of incident neutrons. The results of Yamasaki were later reevaluated by T. Hamada [15].

In 1960, N. Saito *et al.* [16] of the University of Tokyo separated cobalt from two steel samples collected at Hiroshima and Nagasaki and measured the activity of ^{60}Co. The result was presented together with the estimated thermal neutron fluence.

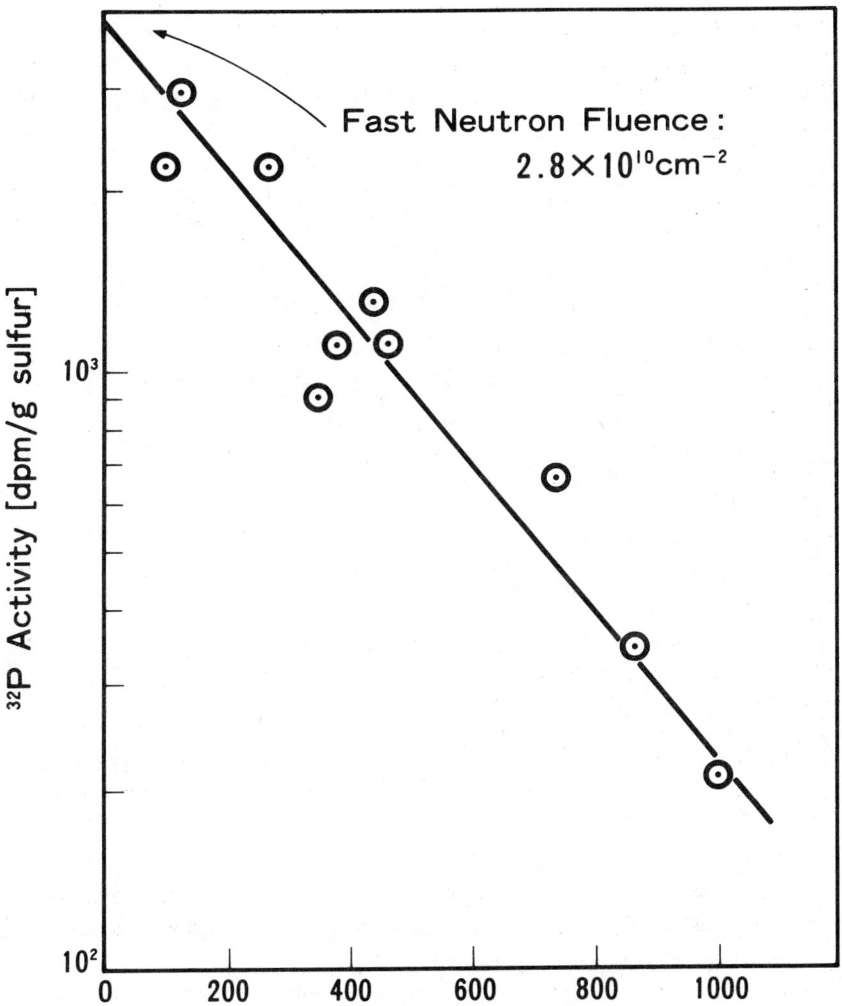

Fig. 4. ^{32}P Activity in Sulfur Samples as a Function of Ground Distance in Hiroshima [12].

In 1965, T. Hashizume and T. Maruyama [17] of the National Institute of Radiological Sciences (NIRS), with the cooperation of the Chemistry Division of their Institute, made extensive measurements of ^{60}Co activity in steel samples in Hiroshima and Nagasaki. In order to measure fast neutrons selectively, they collected steel bar samples embedded in a concrete building at a depth of 8 cm, where the thermal neutron fluence derived from the incident fast neutrons was considered maximum. The cobalt activities were then converted to kerma in tissue by applying a conversion coefficient obtained experimentally with neutrons from HPRR of ORNL and from a van de Graaff generator. The result was presented together with the data of E.N. York in 1957. However, the lack of an appropriate neutron source for calibration makes the validity of the resulting kerma values subject to doubt.

7. Thermoluminescence Measurements

In 1965, Y. Ichikawa *et al.* [18] of Kyoto University and Hashizume *et al.* [17] of NIRS independently made thermoluminescence measurements for gamma-ray dosimetry in Hiroshima and Nagasaki.

Ichikawa collected roof tiles, and the colorless fractions of the powdered samples were subjected to heating. The light output was calibrated by using gamma rays from a ^{60}Co standard source. The results were compared with York's data.

Hashizume used bricks and polished wall tiles attached to concrete buildings. The calibration was made by using a mixed field of x rays from a 6 MeV linac and gamma rays from a ^{60}Co source which is believed to be equivalent to the gamma rays from the bombs in energy distribution. The results, again, were compared with the data of York.

The early works reviewed above, especially those which were carried out within several years following the bombing, are considered to be inadequate. These works, however, provided valuable data for the present reassessment activities.

References

1. Collection of Reports on Investigations of the Atomic Bomb Casualties, vols.1 and 2, Science Council of Japan (1953).
2. Kimura, M. and Tajima, E. (1953). "Epicenter and Size of Fireball of the Hiroshima Bomb," *ibid.* vol.1, p83.
3. Kimura, M., Akutsu, J. and Tajima, E. (1953). "Determination of Intensity of Radiant Heat Rays and Duration of Flash in the Hiroshima Atomic Bomb," *ibid.* vol.1, p88.

4. Watanabe, T., Yamasaki, M., Kojima, B., Nagasaka, S., and Hirayama, K. (1953). "Geological and Petrological Study of Damages Caused by Atomic Bombs in Hiroshima and Nagasaki," *ibid.* vol.1, p143.
5. Tajima, E. (1983). "Estimation of the Hiroshima Bomb Yield and Weather Conditions at the Time of the Bomb," Second U.S.-Japan Workshop for Reassessment of Atomic Bomb Radiation Dosimetry in Hiroshima and Nagasaki, p1, Radiation Effects Research Foundation.
6. Uda, M., Sugahara, Y., and Kita, I. (1953). Meteorological Conditions Related to the Atomic Bomb Explosion in Hiroshima," Collection of Reports on Investigations of the Atomic Bomb Casualties, Science Council of Japan, vol.1, p119.
7. Ishida, T. (1953). "Cloud from the Atomic Bomb Explosion observed from Unzen," *ibid.* vol.1, p139.
8. Yamasaki, F. (1953). "Residual Radioactivity in the West of Hiroshima," *ibid.* vol.1, p25.
9. Shinohara, K., Morita, S., Kawai, N., Yokota, M., and Kohra, K. (1953). "Radioactivity of the Ground in Nagasaki City and Vicinity, II," *ibid.* vol.1, p45.
10. Fujiwara, T. and Takeyama, H. (1953). "Residual Radioactivity around Hiroshima City," *ibid.* vol.1, p75.
11. Miyazaki, Y. and Masuda, T. (1953). "Atomic Bomb Induced Radioactivity in Hiroshima City and Vicinity, II," *ibid.* vol.1, p35.
12. Yamasaki, F. and Sugimoto, A. (1953). "Radioactive ^{32}P in Sulfur in Hiroshima," *ibid.* vol.1, p18.
13. Yamasaki, F. (1959). "Estimation of Radiation Dose due to the Bombing Attack, I," Annual Report of Cooperative Research, Radiation Section, Japanese Ministry of Education, p121.
14. Tajima, E. and Oda, N. (1959). "Estimation of Radiation Dose due to the Bombing Attack, II," *ibid.* p122.
15. Hamada, T. (1983). "^{32}P Activity Induced in Sulfur in Hiroshima: Reevaluation of Data by Yamasaki and Sugimoto," Second U.S.-Japan Workshop for Reassessment of Atomic Bomb Radiation Dosimetry in Hiroshima and Nagasaki, p52, Radiation Effects Research Foundation.
16. Saito, N. (1961). "Estimation of Neutron Fluence of Hiroshima and Nagasaki Atomic Bombs—Radiochemical Estimation," Annual Report of Cooperative Research, Radiation Section, Japanese Ministry of Education, p142.
17. Hashizume, T., Maruyama, T., Shirogai, A., Tanaka, E., Izawa, M., Kawamura, S. and Nagaoka, S. (1967). "Estimation of the Air Dose from the Atomic Bombs in Hiroshima and Nagasaki," Health Physics, 13, 149.
18. Ichikawa, Y., Higashimura, T., and Sidei, T. (1966). "Thermoluminescence Dosimetry of Gamma Rays from Atomic Bombs in Hiroshima and Nagasaki," Health Physics, 12, 395.

Historical Perspectives

William C. Roesch
Richland, WA

In 1981 a reassessment was begun of the doses of ionizing radiation received by the survivors of the atomic bombings of Hiroshima and Nagasaki. That work resulted in a new dosimetry system that is now in use to make these reassessments. This article will sketch the events that led up to the reassessment and how the necessary research was carried out, as seen from the perspective of those in the United States.

RERF

In 1947, the U.S. National Academy of Sciences, at the direction of President Truman, established the Atomic Bomb Casualty Commission (ABCC), with offices in both Hiroshima and Nagasaki, to study the effects of the irradiation of the survivors. The ABCC worked with the Japanese National Institute of Health to this end until 1975. Then Japan and the United States created a cooperative, binational organization, the Radiation Effects Research Foundation (RERF), to take over the work. The two governments share the funding and administration of RERF; the U.S. funds are supplied by the Department of Energy and are made available through the National Academy of Sciences-National Research Council (NAS-NRC). RERF collects, stores and interprets the information about the survivors. The new dosimetry system resulting from the work being described here is installed on the RERF computer at Hiroshima.

T65D Doses

The doses that are being reassessed were determined in 1965. They are called the T65D doses (tentative 1965 doses—"tentative" because the scientists who developed the 1965 dosimetry system felt it would be replaced by a better system).

The 1965 dosimetry system was based on the results of a research program called ICHIBAN that was carried out by the Oak Ridge National Laboratory (ORNL) and on some of the measurements carried out in Japan that were described in the preceding paper by Dr. Hamada. ABCC, with the collaboration of ORNL, worked out a systematic way of applying these data to individual survivors to derive the T65D doses. An excellent description of the ICHIBAN program and the T65D and earlier dosimetry systems is given in a book by J.A. Auxier [1].

In brief, the ICHIBAN program included measurements at weapon tests, on weapon components, at a reactor, and at a high tower on which was mounted, at different times, an unshielded reactor, a large ^{60}Co gamma-ray source, and high-intensity neutron generator. The primary data resulting from these studies were: (1) what we now call the kerma free-in-air (or the free-field kerma), the kerma that the radiation at a point in an arrangement including only the bomb, the atmosphere, and the ground would produce in soft tissue, and (2) measurements made in and around mock Japanese houses to estimate the shielding they provided.

Technicians from ABCC interviewed most of the proximally exposed (within 1,600 m of ground zero at Hiroshima and 2,000 m at Nagasaki) survivors to determine their location at the time of the bombing and how they were shielded from the radiations from the bomb. The T65D doses were calculated from these data.

NCRP

The reassessment was the result of several streams of events. One of these involved the National Council on Radiation Protection and Measurements (NCRP). In 1976, H.H. Rossi presented a talk at the Council's annual meeting in which he combined new calculations of depth doses by Jones [2] with T65D dose estimates and concluded that a radiation worker continually exposed to neutrons at the maximum permissible level recommended by the NCRP had a probability of developing leukemia that was several times greater than that of a person receiving the maximum permissible level of x or gamma rays. Rossi recommended that the NCRP reduce its permissible doses for neutrons by an order of magnitude. The subject of his talk was later developed further in an article with C.W. Mays [3].

One of the first responses of the NCRP was to set up a Task Group, under the chairmanship of H.O. Wyckoff, to estimate the accuracy of the T65D doses. The Task Group began its work in 1976. After considerable study, it concluded that the material in the literature and in published laboratory reports was insufficient for a determination of the accuracy of the T65D system.

Some of the problems with the available material are illustrated in Figures 1 & 2. Figure 1 shows, as a function of the distance from the center of the bomb, the product of the kerma in tissue due to gamma rays and the square of the distance, all divided by the yield of the bomb. This quantity is shown for Hiroshima and Nagasaki, with kermas from the T65D system, and for four other bombs for which data are available in the open literature [4,5]. Multiplying by the square of the distance compensates for the inverse-square effect; the plot indicates the remaining attenuation of the radiation. Notice the difference in the attenuation rate of the gamma rays at Hiroshima and Nagasaki; this difference was not satisfactorily explained.

The yield is the amount of energy released in the explosion (usually given in the kilotons of TNT that would give an equivalent energy release). The yield is indicative of the amount of radiation initially released in the explosion. Other things being equal, dividing by the yield should make the different curves coincide. Clearly other things are not equal: none of the other bombs coincide and both Hiroshima and Nagasaki lie considerably below the other bombs—as much as an order of magnitude below. Figure 2 shows the data for neutrons; again there are very large differences. These large differences are due to differences in attenuation of the radiation in the structures of the bombs, which were very different. The information needed to understand these differences is classified secret because of its importance to the national security of the United States.

The Task Group recommended that a person with the proper security clearance should complete the review using both classified and unclassified sources of information. They hoped that after this was done enough information could be made publicly available to provide a secure foundation for radiation recommendations based on the data concerning the bomb survivors.

The Energy Research and Development Administration (now DOE) responded by funding G.D. Kerr of ORNL to start such a study in 1979. Kerr included not only classified data in his study but began incorporating much new data that had not been applied yet to bomb dosimetry. Use of the new data led to other investigators being drawn

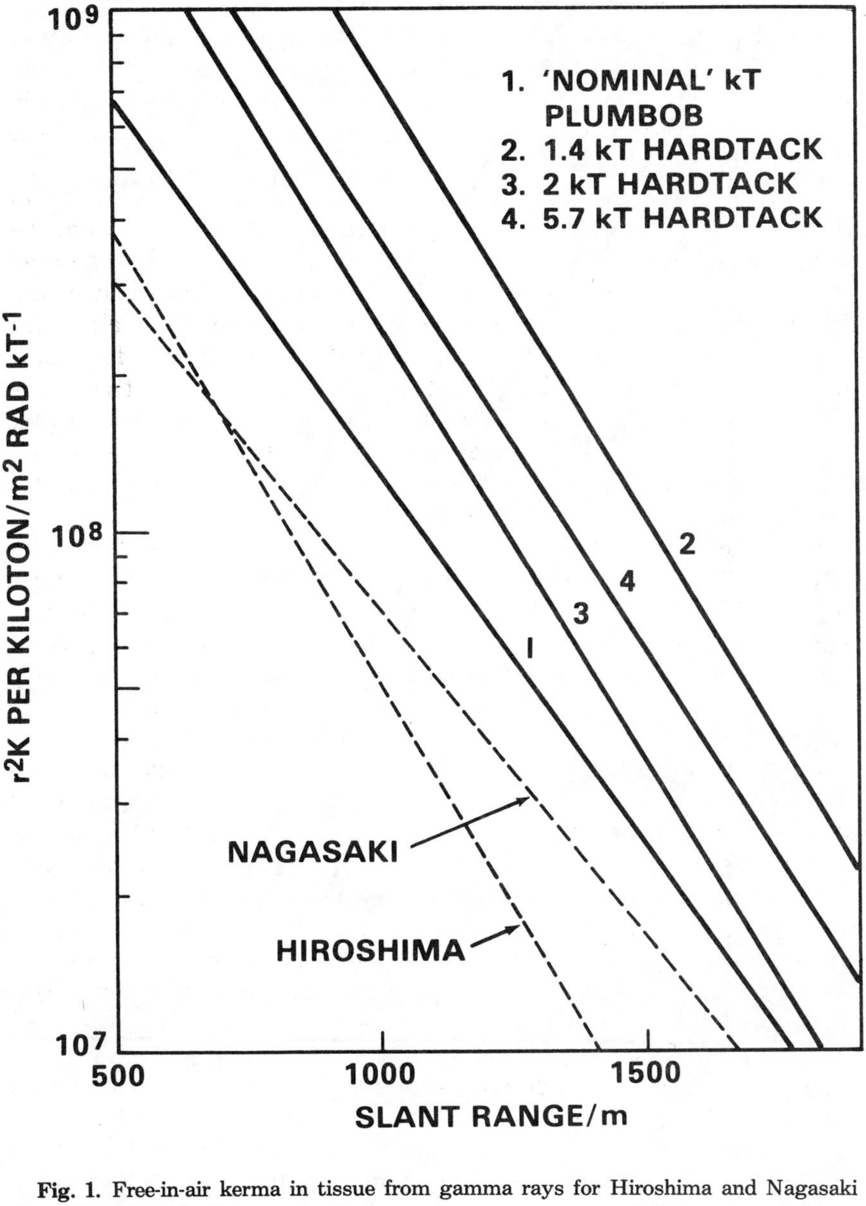

Fig. 1. Free-in-air kerma in tissue from gamma rays for Hiroshima and Nagasaki (from T65D) and for four weapon tests.

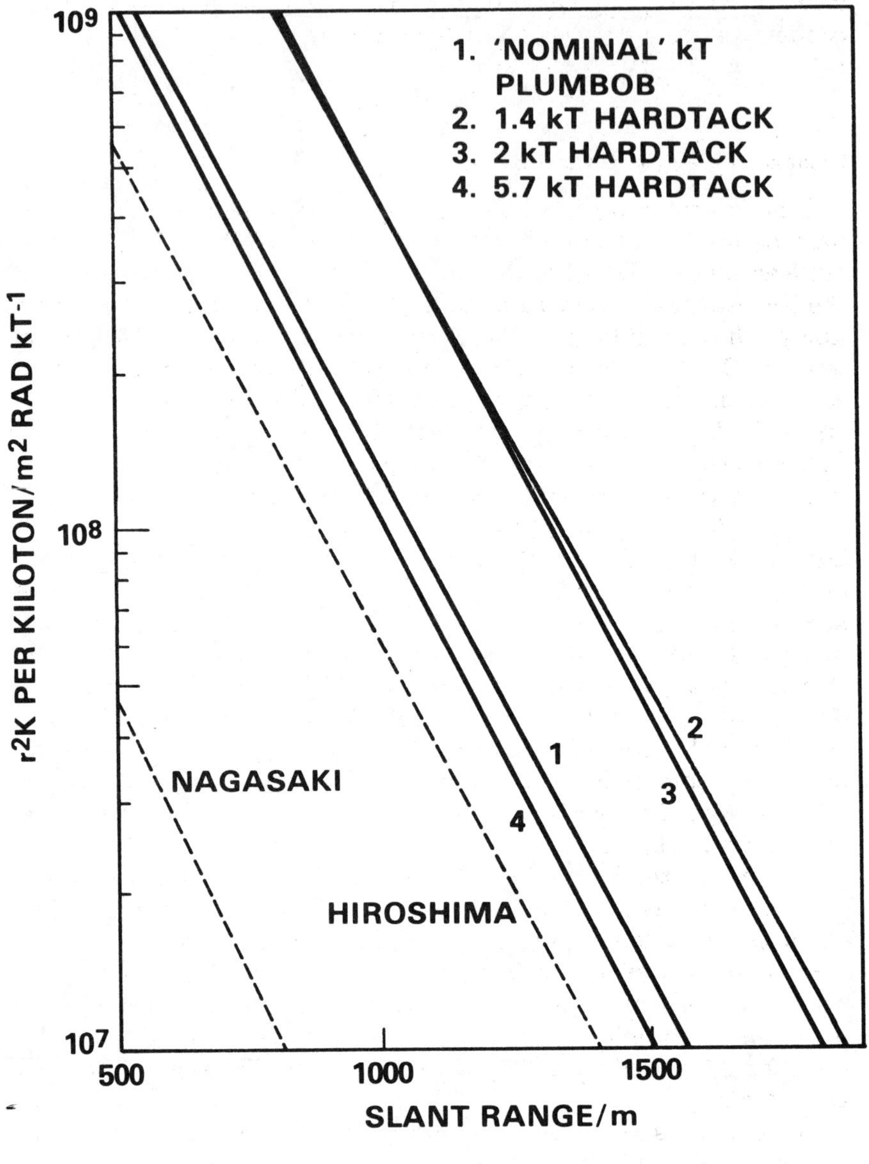

Fig. 2. Free-in-air kerma in tissue from neutrons for Hiroshima and Nagasaki (from T65D) and for four weapon tests.

into the review. They came to the conclusion that the T65D system contained errors and needed further research.

Leakage and Transport Calculations

Other events were also showing that a reassessment was needed. As part of their program of studying nuclear weapons, the Defense Nuclear Agency (DNA) of the U.S. Department of Defense had asked the Los Alamos National Laboratory (LANL) to calculate how radiation produced inside exploding bombs was attenuated in leaving the weapon. C.P. Knowles of R & D Associates (RDA) requested that the calculations also be performed for the Hiroshima and Nagasaki bombs. In 1976, W.E. Preeg of LANL sent Knowles his calculations of the leakage spectra, the spectra immediately outside the two bombs. The letter was widely distributed and has since been published [6].

The availability of the spectra of the radiations emerging from the bombs led a number of people to apply available transport codes to the calculation of the penetration of these radiations through the atmosphere. Preeg himself used the HEART code and found neutron kermas for Hiroshima that were much smaller than those given by T65D. Similar results were found in 1977 by J.V. Pace of ORNL, using the DOT-3 code, and by D.C. Kaul and R. Jarka of Science Applications International Corporation, using the ATR-4 code. G.D. Kerr of ORNL and J. Marcum of RDA, in independent reviews of these results for DNA, recognized their importance to the survivor dosimetry and recommended further research.

Scientist at the Lawrence Livermore National Laboratory (LLNL) were quite concerned about the recommendations of Rossi and Mays about permissible neutron doses; they began, independently, a thorough study of survivor dosimetry. Loewe and Mendelsohn [7] reported neutron doses considerably lower than the T65D values, and Straume and Dobson [8] applied their results to published survivor data to judge the effect of revised dosimetry data on the interpretation of the radiation effects.

Other aspects of weapons radiation calculations also came under scrutiny and were realized to need improvement. For example, Pace [9] found that the considerable reduction in the average energy of the neutrons escaping the bombs resulted in their transport being quite sensitive to the humidity of the air. And Scott [10] showed that previous estimates of the gamma radiation from fission products in the fireball were not as accurate as had been thought.

The Reassessment

In the spring of 1981, the NCRP cosponsored with the North American Late Effects Group a meeting on survivor dosimetry at the annual meeting of the Radiation Research Society. The attendant publicity in both the scientific and lay press brought the situation to the attention of interested scientists and of political representatives of the Japanese and United States governments. Both scientific and social forces began to press for a reinvestigation of the doses the survivors had received.

In the fall of 1981, DOE held a symposium [11] to review what was known about survivor dosimetry and to determine what further work needed to be done. Afterward, they set up a Working Group on A-Bomb Dosimetry Reassessment to carry out the U.S. program. Dr. R.F. Christy of the California Institute of Technology was appointed its chairman. DOE also asked the National Academy of Sciences to form an oversight committee to the Working Group. The Academy appointed a Panel on Reassessment of Atomic Bomb Dosimetry with Dr. F. Seitz as chairman.

In 1981 the Japanese Ministry of Health organized a research team similar to the DOE Working Group; it was chaired by Dr. T. Hashizume. Later it was reorganized into two groups, the Review committee to Evaluate the Radiation dose from the Atomic Bombs (the Japanese Senior Dosimetry Committee) and the Japanese Dosimetry Working Group, both under the chairmanship of Dr. E. Tajima.

The Japanese and U.S. groups worked out a joint research program coordinated through RERF. Four joint workshops were held. The first two were scientific workshops and published proceedings (Nagasaki, February, 1983, [12]); Hiroshima, November, 1983, [13]) as well as providing opportunities for joint planning. Agreement was reached that the final "product" of the research program was to be a computer program embodying the best of all the research results. The program would be provided to RERF for use in their analysis of the survivor data.

By early 1985, work had progressed to the point that a joint workshop at Pasadena in March was devoted to the review of rough drafts of manuscripts for a final report on the survivor dosimetry study. A preliminary version of the computer program, dubbed DS84 (dosimetry system 1984), had been tried and some preliminary results were available.

The last joint meeting was held at Hiroshima in March 1986 to review the thermoluminescence dosimetry and neutron measurements programs. But the principal purpose of the meeting was to give final agreement to the computer program, DS86 (dosimetry system 1986).

The program was formally presented to RERF, which officially adopted it a short time later.

A report that describes the research on survivor dosimetry and the computer program DS86 was released by RERF [14].

DS86

The mnemonic T65D included a "T" (for tentative or temporary) because it was believed that further work might alter parts of that dosimetry system. The participants in the research that led to the new 1986 system also feel that future work will lead to changes in DS86. Some of these changes will just be additions to enable treating more classes of survivors. Others might be fundamental changes in how doses are calculated—these might, for example, reflect improvements following resolution of current problems in comparing calculations with measurements that you will hear about in other papers at this meeting. The general belief at the present time is that any such fundamental changes will result in only small changes in the doses. The DS86 system was put together in such a way that any such changes can be incorporated easily, without major alterations to the system. In the future, improvements to the dosimetry system must by expected and accepted in the same spirit that changes and additions to the epidemiological data base are accepted.

References

1. Auxier, J.A. (1977). *Ichiban: Radiation Dosimetry for the Survivors of the Bombings of Hiroshima and Nagasaki*, TID-27080 (Department of Energy, Washington).
2. Jones, T.D. (1975). "Radiation insult to the active bone marrow as predicted by a method of CHORDS," manuscript; later (1976) issued as a report, ORNL/TM-5337, by the Oak Ridge National Laboratory, Oak Ridge, TN.
3. Rossi, H.H. and Mays, C.W. (1978). "Leukemia risk from neutrons," Health Physics 34, 475.
4. Ritchie, R.H. and Hurst, G.S. (1959). "Penetration of weapons radiation: application to the Hiroshima-Nagasaki studies," Health Physics 1, 390.
5. Auxier, J.A., Cheka, J.S., and Sanders, F.W. (1961). *Attenuation of Weapons Radiation: Application to Japanese Houses*, report WT-1725 (Oak Ridge National Laboratory, Oak Ridge, TN).

6. Preeg, W.E. (1976). Letter to C.P. Knowles, R & D Associates, published as an appendix to an article by Whalen, in *Reevaluations of Dosimetric Factors, Hiroshima and Nagasaki,* V.P. Bond and J.W. Thiessen, Eds., pp. 125-130, report CONF-810928 (Department of Energy, Washington).
7. Loewe, W.E. and Mendelsohn, E. (1981). "Revised dose estimates at Hiroshima and Nagasaki," Health Physics 41, 663.
8. Straume, T. and Dobson, R.L. (1981). "Implications of new Hiroshima and Nagasaki dose estimates: cancer risks and neutron RBE," Health Physics 41, 666.
9. Pace, J.V. (1979). Letter to J. Marcum, R & D Associates. (Oak Ridge National Laboratory, Oak Ridge, TN).
10. Scott, W.H. (1981). Letter to G.D. Kerr, Oak Ridge National Laboratory (Science Applications International Corporation, La Jolla, CA).
11. *Reevaluations of Dosimetric Factors, Hiroshima and Nagasaki,* V.P. Bond and J.W. Thiessen, Eds., CONF-810928 (Department of Energy, Washington).
12. *U.S. Japan Joint Workshop for Reassessment of Atomic Bomb Radiation Dosimetry in Hiroshima and Nagasaki* (1983) (Radiation Effects Research Foundation, Hiroshima).
13. *Second U.S. Japan Joint Workshop for Reassessment of Atomic Bomb Radiation Dosimetry in Hiroshima and Nagasaki* (1984) (Radiation Effects Research Foundation, Hiroshima).
14. *US-Japan Joint Reassessment of Atomic Bomb Radiation Dosimetry in Hiroshima and Nagasaki, Final Report (1987),* W.C. Roesch, Ed., two volumes (Radiation Effects Research Foundation, Hiroshima).

Overview of the New Dosimetry: The Physical Basis

Robert F. Christy
California Institute of Technology
Pasadena, California

(Manuscript not received for Publication)

Discussion

H. FRIEDELL: I came here primarily to hear about this intriguing question, but since you introduced the early history of measurements, I think I have something to add that may be useful to the people that are studying this question. First, let me identify the capacity in which I participated. In 1945, I was Lt. Col. in the Corp of Engineers. I was then deputy chief of the medical division of the Manhattan Corp of Engineers under Stafford Warren (its executive officer). Immediately after the explosion, units were drawn to enter Hiroshima and Nagasaki and take measurements of the radiation and its biological effects primarily with 2 objectives: (1) to corroborate the fact that indeed biological effects had occurred, and (2) to see whether American personnel would occupy that area. We, therefore, were immediately assembled. The units entered Hiroshima and Nagasaki. I led the unit into Hiroshima and we spent about a week making the measurements. We were primarily interested in radiation levels at the time, for personnel that might be there. We used ionization chambers and also some geiger counters, criss crossing all of the area many people know, between two mountain ridges with a river coming down which divides into various tributaries. In any event, we made these measurements and recorded an isodose curve in which we identified the radiation levels approximately in the order of 10 to 15 percent of background. We arrived there 3 to 4 weeks after the explosion because the typhoons, that had been occurring at that time, delayed our entrance. In any event, we made these measurements, and recorded them, and as I recall, our reports, which exist somewhere in the records of the Manhattan Corp of Engineers, state that we found radiation levels of approximately 10 to 15 percent of background; a little higher toward the mountains. The isodose curves, as I recall, were higher nearer the point of the epicenter and receded from that area. We had an inadvertent calibration of our instruments in the fact that we found some radium which had been lost by the Japanese. We did identify radiation levels which were very low, and our final reports indicated that occupation by American persons was possible.

C. LARSON: I'd like to ask a question of Dr. Christy with regard to the yield. I believe at the time of the dropping of the Hiroshima bomb there was Alvarez who made measurements as to the yield, and my recollection is that the yield that he determined was closer to 13 rather than 16. And I wonder if you care to comment on the role of Alvarez's measurements.

R. CHRISTY: The measurements that are usually recorded, the blast measurements, are those of Penney and Associates, they certainly determined the yield on the order of 12 kT. On the other hand, the model they used was an air blast over a very clean surface and there has been some concern that that model might be untrue. There were measurements made, or attempted to be made by the US group from cannisters having air pressure gauges or microphones, that were dropped in association with the bombs, and then the results from those were recorded in some of the aircraft. Those measurements were not quite fully satisfactory and there is still work going on trying to analyze those data. It has not been entirely completed, so there might be changes in the results of the analysis of those data. It is true that there have been differing estimates over the course of time in all of these things. This is probably one of the most difficult quantities to ascertain, the yield of the Hiroshima bomb, because there are no really good solid measurements, and the best measures that one has been able to put together didn't really agree with each other. It was for that reason that it was of the greatest importance to try to measure directly at some distance, the gamma rays and, if possible, the neutrons, so that one did not have to rely on a lengthy calculation starting with the yield of the bomb.

C. LARSON: Thank you very much. I think that answers the question. In view of the fact that previous people have mentioned some historical information, I should like to point out that the 20 kT which President Truman had given was purely one of those gross estimates, and when Alvarez got these readings, no one thought it very prudent to release the 12 to 13 kT measure.

A. BRODSKY: I was looking at some of the points on the error distribution plots and I noticed that in some of the cases, more of the points were up toward the 1.5, from zero, than down to the -0.5 level, which kind of indicates that the error distributions in some of these cases might be more than normal, and really help not to underestimate the range of the error if, as we went along, the error distributions or the data were plotted on probability vs. log-scale paper to see whether they are normal or log normal because then you might way that the error, instead of ± 50 percent, might be times and divided by 2, or times and divided by 1.5. It looked like it was probably true in some of the cases.

R. CHRISTY: I was not trying to comment on the distribution of the errors. I did comment that the indications from those measurements were that the calculated curve might be a little low in the region that you mentioned. You understand that the DS86 is, in its entirety, basically a calculated system, that starts with the bomb going off, calculates the emission of radiations that are transported, and so forth. It is, however, validated by various measurements and I tried to show how good that validation was, or how not good, in various places. Where the measurements clearly indicate that the dosimetry will need revision, then we will certainly have to make a revision which comes closer to the measurements. But at present, the system is a calculated system, which is however, compared to measurement wherever feasible.

C. MAYS: Question for Dr. Christy. That last slide you showed indicated that at large distances, the sulphur measurements indicated that there might have been more neutrons at Hiroshima than have been calculated. And I wish you'd comment on whether you believe the actual doses could in fact be 3-fold higher from neutrons at Hiroshima at large distances, than calculated.

R. CHRISTY: There were a few measurements of sulphur neutrons from insulator activation which went out to the order of 1 km or more, and those measurements that were easily reported were well above the calculations, as you may have noted. However, there were some other measurements that were not so readily reported that were below. The reasons they weren't reported was that the measurements were below background and therefore, after you subtract background, you had a negative measurement. It's not considered in good scientific taste, to show a negative measurement and therefore, you can show those properly only as I did on the previous curve, where the measurements, including background, were plotted. So the reasons why you saw some points there that were quite high was because when you're measuring below background, you'll find some points which are above and some points below. And the points below background weren't shown. As to the question you raised about whether or not there could be a major error in the neutron at some distance, I tried to point out, and there are of course, other ways of verifying it, that in other cases, neutrons have been measured out to great distances of 2 km or more (bomb tests) and the agreement with calculations following these same models has been excellent. Therefore, there is basically no room for a major error in how neutrons are propagated. It has been possible to verify that the calculations propagate the neutrons properly. The only question could arise as to whether the source spectrum of the neutrons is adequate. Now, I showed that certainly neutrons on the order of 3 MeV must be approximately correct because they agree with the sulphur activation which requires approximately 3 MeV neutrons. We don't have a direct measure of the intensity of 1 to 2 MeV neutrons. If it is conceivable that there is somewhat greater source intensity there than we're calculating, then that might contribute to some added flux at larger distances, although the purpose of the critical test with the mockup I showed was to see whether the system that we were using was able to calculate neutrons over the whole spectrum with reasonable accuracy. And it seemed that it could. But I have to admit that there are some holes in this and that we have not been able to reproduce the thermal neutron activation at about 1.2 km yet. I don't think that this will result in any corrections even as big as a factor of 2. But that's just based on my personal estimation, I can't quite prove it to you.

W. SINCLAIR: Dr. Christy, in the case of Hiroshima and the gamma rays, the DS86 is not only above the T65 but the slope is less. And in the case of Nagasaki, the DS86 is not only below the T65 but the slope is steeper. Is there a reason for those differences in slope that I missed?

D. KAUL: The interesting thing about T65 was that the neutron relaxation lengths at both cities were taken to be the same. The gamma-ray relaxation lengths at both cities were different. DS86 is exactly the opposite of that. The 2 neutron relaxation lengths were different and the 2 gamma-ray relaxation lengths were almost the same. Now you must remember that T65 was an empirical system. They worked with what they had. What they had was a great deal of modern neutron source measurements, which, as you saw from the slide that was shown earlier, I believe by Bill Roesch, that nothing there indicated that they should see anything in the way of neutron slopes but what they saw, and so they assigned constant relaxation lengths to the neutrons. Now what they had for gamma rays, they had from a device that was similar to the Nagasaki shot, not anything like the Hiroshima shot. They had Ranger Fox, a shot which is virtually identical to the Nagasaki device, except for the fissile material inside it. But they didn't have anything like Hiroshima. So I can only conjecture that when they took that empirical data, they looked at that Hiroshima device and said well there's something different here. It's a heavy steel device, and so there's going to be less energetic gamma radiation coming out of it. So when they created the T65 they actually discounted the gamma-ray source and its energy out of the little-boy Hiroshima device. As a result, at Hiroshima, T65D is lower than the DS86 because in DS86 the crux is that the gamma rays come from 2 places. They come from neutron generated gamma rays in the air and they come from delayed gamma rays that come out after the device is basically evaporated and can no longer attenuate the gamma rays. The same thing is true in the case of the Nagasaki device. As a result, the modern calculations show that the relaxation lengths of the gamma rays of the 2 devices are virtually identical. Whereas with the empiricisms that were used in T65, it was argued, that they were not.

R. CHRISTY: I wonder if I might make one addition to my comments. I neglected to comment on the role of the National Academy and I would like to, at this point, thank Seymour Jablon for his role as kind of the overall wiseman behind our efforts; to Bill Ellett, for his efforts as the man with the whip who kept us going; and to Wayne Lowder whose perceptive distribution of money made the US efforts possible.

W. SINCLAIR: Could I just ask Dr. Kaul if his answer means that the slopes of the gamma rays in DS86 at Hiroshima and Nagasaki are exactly the same.

D. KAUL: They're very similar compared to T65. There are differences. There are differences because the Nagasaki device was a plutonium device and therefore, had fewer delayed gamma rays out per kT or per fission as did the Hiroshima device. The Nagasaki device actually leaked a few more neutrons per kT than the Hiroshima device. So, the result is that there are small differences and you saw different crossovers in Dr. Christy's slide between the various components in the two. But basically, since the gamma rays come from the same basic mechanisms in both bombs, and almost in the same proportions you would expect that the slopes would be very similar, and they are.

Scientific Session

Special Aspects of Dosimetry

Frederick Seitz,
Chairman

Introduction

Frederick Seitz
Rockefeller University

When the Department of Energy created the illustrious team that was led by Prof. Christy, it also asked the Academy to form a Committee of experts to cooperate with him. I had the privilege of chairing that Committee of very fine people. Perhaps Prof. Christy had some misgivings to begin with but I hope he found the work of the Committee more helpful than not. The fact that he thanked the staff so profusely is an indication that the Committee and apparatus that went with it was most useful, in fact, as Chairman of that Committee, I'm delighted to join with him in thanking Mr. Jablon and Dr. Ellett for everything they did to keep the story moving. One thing that came out of the work of the Committee of course, was that we found in it a wonderful person to pull together all fragments of the final Report — namely Dr. Roesch, who you heard this morning. I might also say that another wonderful thing that went with working with both that Committee and with Prof. Christy and his group is the association with our colleagues in Japan. I'm particularly grateful, and I would like to thank Dr. Shigematsu and Dr. Tajima for the many occasions in which we met and worked together and for all the cooperation that we obtained. Many people worked in the study, as you know, under Prof. Christy's direction, counterparts in Japan. We will now hear from a few of them in connection with their own work which added to the whole and will appear in the Report.

The Use of Thermoluminescence

Edwin H. Haskell
University of Utah

I will summarize this morning the results of a multinational effort by thermoluminescence (TL) laboratories in Japan, the United States and England [1]. The individuals and laboratories responsible for the TL measurements of bricks and tiles at Hiroshima and Nagasaki and the comparison of those measurements with theory are listed in Figure 1.

Thermoluminescence is synonymous with personnel dosimetry for many people who routinely work with radiation. TL dosimeters (TLD) provide measurements of the dose which accumulates from the time they are annealed by heating to the time they are measured in the laboratory. The measurement process involves measuring the light output form the TLD's as they are heated under controlled conditions. The amount of light emitted at characteristic temperatures is a function of the dose which was received.

Thermoluminescence is not unique to commercially prepared dosimeters (Figure 2) but is a property of a great many naturally occurring materials. Common sand, SiO_2 can be an excellent TL material. Its main drawback is that its annealing history is unknown. For sand or any other TL material to be useful as a natural contemporary dosimeter, it must have been annealed at a relatively recent date. Common materials which best meet this requirement are ceramics such as bricks and tiles for which the date of manufacture can be determined.

Physicists working in archaeology have used thermoluminescence to date ancient pottery samples for over two decades. By measuring the total dose of radiation in a sample and then dividing by the annual dose-rate, the age of the sample can be estimated (Figure 3). It has been the archaeological use of TL which has resulted in many of the technique refinements used in the present study [2].

TL Measurements

T. Maruyama, Y. Kumamoto
National Institute of Radiological Sciences (NIRS)
Japan

Y. Ichikawa, T. Nagatomo
Nara University of Education (NUE)
Japan

M. Hoshi
Hiroshima University (HU)
Japan

M. Aitken, D. Stoneham, J. Huxtable
Oxford University (OXF)
U.K.

I. Bailiff
Durham University (DUR)
U.K.

E. Haskell, P. Kaipa
University of Utah (UU)
U.S.A.

Comparison with Theory

D. Kaul, S. Egbert, T. Kuhn, J. Roberts
Science Applications International Corporation
U.S.A.

Fig. 1. Participants in thermoluminescence measurements and comparisons with theory

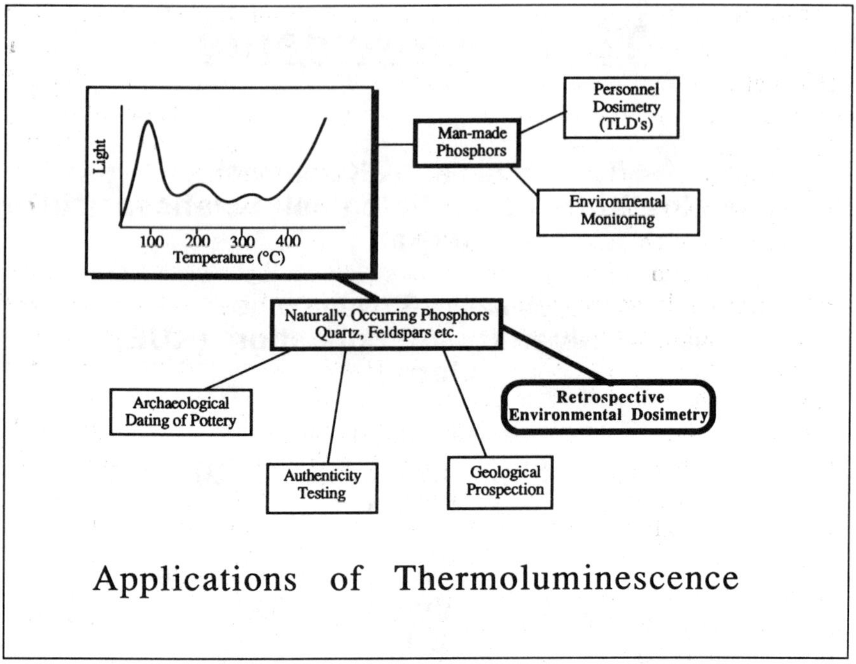

Fig. 2. Applications of thermoluminescence

Age Determination Using Thermoluminescence Techniques

$$\text{Age} = \frac{\text{Natural Dose}}{\text{Annual Dose-Rate}}$$

Natural Dose = Dose accrued since last firing of the sample.

Annual Dose-rate = Dose-rate to sample grains from alpha, beta, gamma and cosmic radiation. Attenuation and efficiency factors must be included. Dose-rate is assumed constant.

Fig. 3. Age determination using thermoluminescence techniques

But the dosimetry aspects of the technique preceded the archaeological uses (Figure 4). In the mid-1960's Dr. Hashizume and his group at the National Institute of Radiological Sciences [3] and Dr. Ichikawa and his group at Kyoto University [4] applied the technique to tiles in Hiroshima and Nagasaki to provide direct measurements of gamma-ray doses at those cities. Because these measurements were close in to the hypocenters, the doses which could be measured were large and not of biological significance to survivors.

Since the time of those early measurements TL techniques on natural materials have evolved to the point where measurements at dose levels of biological interest are feasible. As a result, it was possible to make direct measurements of doses received in Hiroshima and Nagasaki from the A-bomb detonations at distances in excess of 1 km.

Over 40 years have passed since the bombings, so the first question to be answered was "are there any sampling sites in Hiroshima and Nagasaki with intact bricks or tiles which have not been affected by fires." Although both cities have been almost totally rebuilt, there were several sites which remained largely as they had been at the time of the bombings. It was through the work of Dr. Hoshi at Hiroshima University and members of the Radiation Effects Research Foundation that samples were identified and obtained.

A-bomb Dosimetry Using Thermoluminescence Techniques

A-bomb Gamma-ray Dose = Accrued Dose − (Age x Annual Dose-Rate)

Accrued Dose	=	Total dose to the sample since firing from all sources
Age	=	Time from firing of sample to analysis
Annual Dose-Rate	=	Dose-rate to sample grains from alpha, beta, gamma and cosmic radiation. Attenuation and efficiency factors must be included. Dose-rate is assumed constant.

Fig. 4. A-bomb dosimetry using thermoluminescence techniques

In Hiroshima, the primary sites were located southeast of the hypocenter (Figure 5). The Faculty of Sciences Building at Hiroshima University (Figure 6) located between 1,380 and 1,460 meters from the

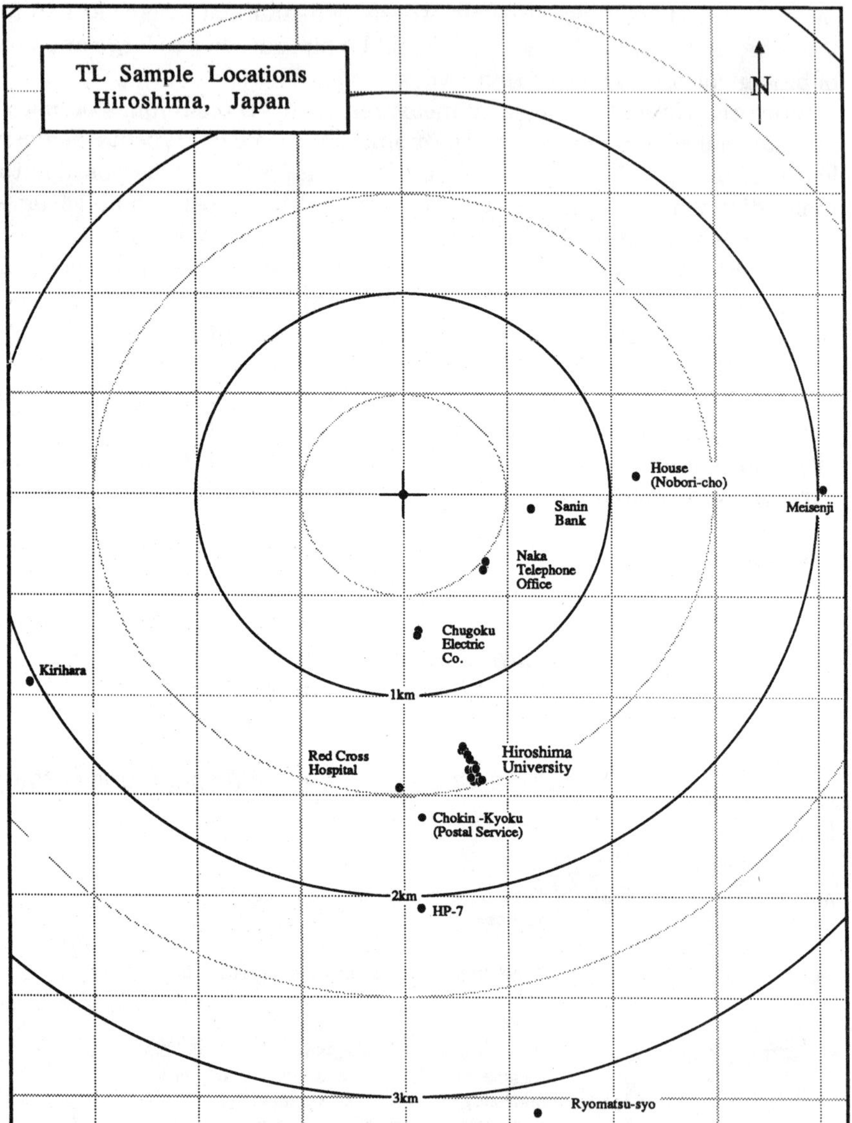

Fig. 5. TL sample locations, Hiroshima, Japan

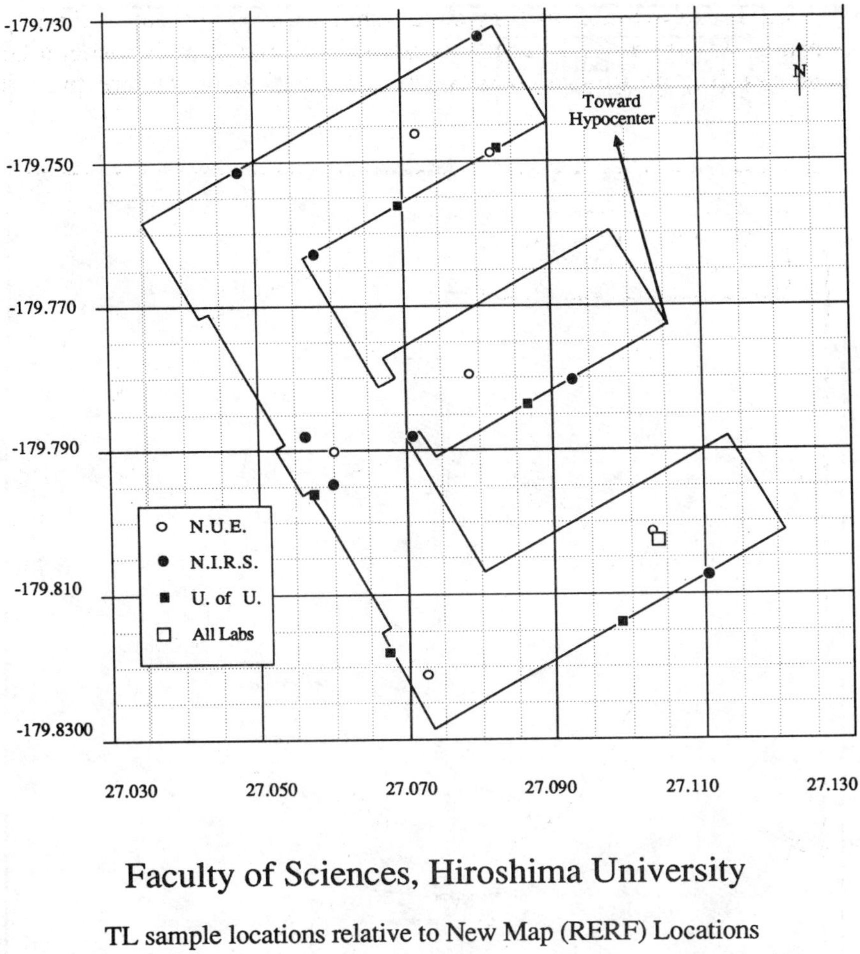

Fig. 6. TL sample locations, Faculty of Sciences Building, Hiroshima University

hypocenter provided the sole material for several of the laboratories. The surface of the building was composed almost entirely of ceramic tiles. In addition to the samples from the Faculty of Sciences Building which were analyzed by all of the laboratories, samples were obtained from 8 other structures and analyzed by the laboratories at NIRS and NUE.

Finding a suitable site in Nagasaki with intact samples was more of a problem (Figure 7). Fortunately, a brick wall which had been directly exposed to the blast was found at a distance of 1,437 meters from the

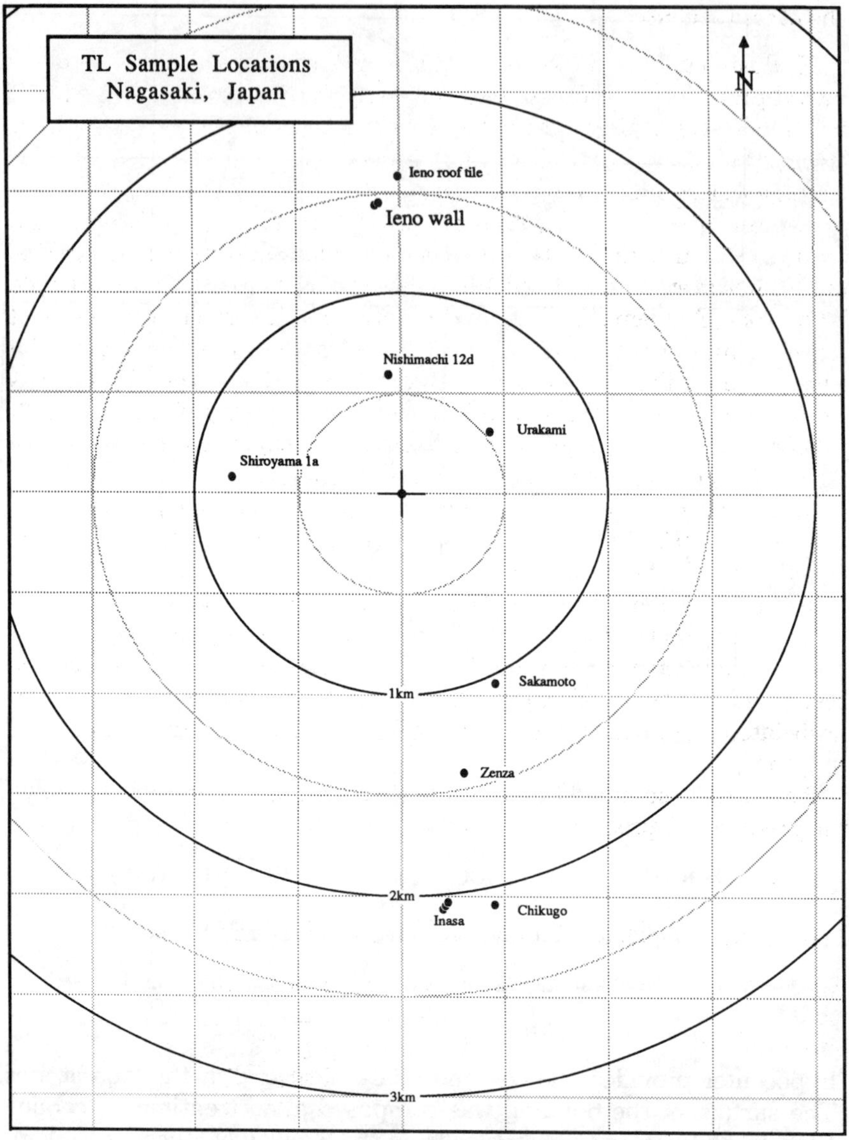

Fig. 7. TL sample locations, Nagasaki, Japan

hypocenter and provided material for each laboratory. As with Hiroshima sites, several additional sites in Nagasaki were analyzed by the NIRS laboratory.

The Measurements

TL dosimetry for personnel monitoring is often a routine, fully automated procedure. Unfortunately the analysis of environmental materials is less straightforward. Since TL properties of samples vary from location to location and tile to tile it is necessary to test the samples for characteristics which may lead to erroneous dose estimates. In addition measurements of annual dose rate from natural sources of radiation are required. Natural dose-rate was measured with several techniques (Figure 8), however, many of the samples were measured by only a single laboratory. Agreement of approximately 14% (1 sigma) was obtained for beta measurements from four of the laboratories measuring the same type of tile. Representative internal beta doses from 3 tile types from the 52-year-old Faculty of Sciences Building ranged from 16 to 19 cGy in quartz. Gamma plus cosmic-ray dose rates measured by gamma TLD averaged approximately .01cGy per year ± 27%.

In general, two independent types of TL analyses were performed on the samples (Figure 9), traditional high temperature TL, and the so-called pre-dose TL technique. The pre-dose technique was used exclusively by the University of Utah while both the predose and high temperature techniques were used on one or more samples by each of the other laboratories. No significant differences were seen between the results of the two techniques. Each method of analysis has its own associated diagnostic tests which are used to rule out sample anomalies (Figure 9).

Tests were also made to check for effects which could be induced by the preparation procedures or could be unique to the A-bomb situation

Measurement of Internal Beta Dose-rate

NUE	Beta TLD
NIRS	Beta TLD
UU	Beta TLD
DUR	Beta TLD
OXF	Alpha Counting

Measurement of Natural Gamma + Cosmic Dose-rate

NUE	Gamma TLD
NIRS	Gamma TLD

Fig. 8. Techniques used for natural background dosimetry

High Temperature TL Analysis

Thermal and non-thermal fading

Supralinearity

Heating effects

Dose-dependent heating effects

Pre-dose TL Analysis

Changes in initial sensitivity

Temperature of maximum sensitization

Saturation

Changes in degree of sensitization following first heating

Fig. 9. Thermoluminescence techniques and associated tests

such as neutron effects and dose-rate effects (Figure 10). No effects on dose estimates were seen from neutron-irradiated samples or for dose rates varying from 0.8 Gy per minute to 30,000 Gy per minute (Figure 11). One test on sample preparation procedures did reveal an effect associated with chemical etching of samples high in sodium feldspar.

A number of interlaboratory comparisons (Figure 12) were also performed, including checks on interlaboratory calibration, ability of the techniques to measure known doses of radiation, and precision obtainable between laboratories on measurements of similar samples.

The interlaboratory calibration used a TL phosphor, magnesium orthosilicate (Mg_2SiO_4:Tb, Kasei Optonix, Tokyo) with an energy response similar to quartz. The phosphor was irradiated at the National Bureau of Standards with a C0-60 source and used to compare the calibrations of the sources used at each laboratory. The results (Figure 13) were used to adjust the final dose estimates from each laboratory.

Each laboratory also measured the dose received by an annealed Nagasaki sample also irradiated at the National Bureau of Standards by

Tests for Miscellaneous Effects

1. Neutron Effects
2. Dose-Rate Effects
3. Preparation Effects

Fig. 10. Tests for miscellaneous TL effects

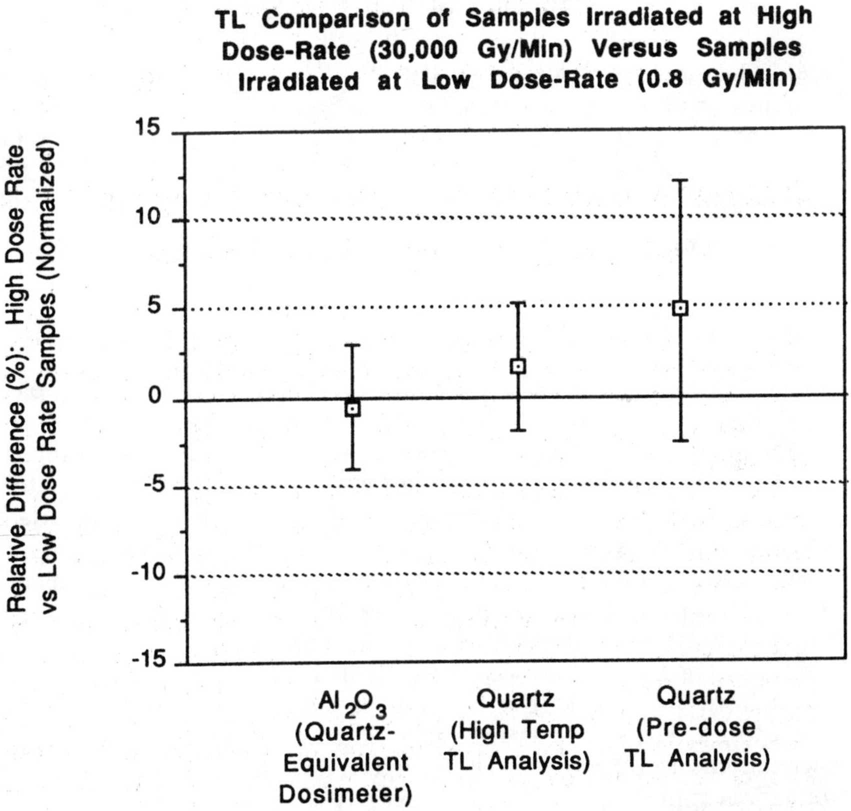

Fig. 11. TL comparison of samples irradiated at high dose-rate (30,000 Gy/min) versus samples irradiated at low dose-rate (0.8 Gy/min)

Interlaboratory Comparisons

1. Interlaboratory Source Calibration Using NBS-Irradiated Phosphor
2. Interlaboratory Comparison of NBS-Irradiated Nagasaki Sample
3. Interlaboratory Comparison of Hiroshima Tile Measurements
4. Interlaboratory Comparison of Nagasaki Brick Measurements

Fig. 12. Interlaboratory comparisons

Calibration Factors Based on Measurement of NBS-Irradiated Phosphor

Nara University of Education (NUE)	0.98
National Institute of Radiological Sciences	1.00
University of Utah	1.11
Oxford University	1.06
Durham University	1.09

Fig. 13. Calibration factors based on measurement of NBS-irradiated Mg_2SiO_4

were similar to those used for the other Hiroshima/Nagasaki samples. Even though the quantity of quartz in the sample supplied to the laboratories was less than that used for analysis of the other samples, the results were encouraging with the between-laboratory means matching applied doses to within 1 cGy at all dose levels. The standard deviation at the highest dose level was approximately ± 13%.

Perhaps the most direct tests on the consistency of measurements from the various laboratories were the intercomparison studies performed on actual samples taken from the two cities.

At the Faculty of Sciences building in Hiroshima, floor tiles were taken from the roof of the building and distributed to each of the laboratories. The samples were analyzed using one or more methods by each laboratory (Figure 15). The standard deviation between all laboratories was approximately 20%, due largely to the inclusion of one value which appears to be an outlier. When the comparison is restricted to the other four laboratories, that value drops to better than 5%.

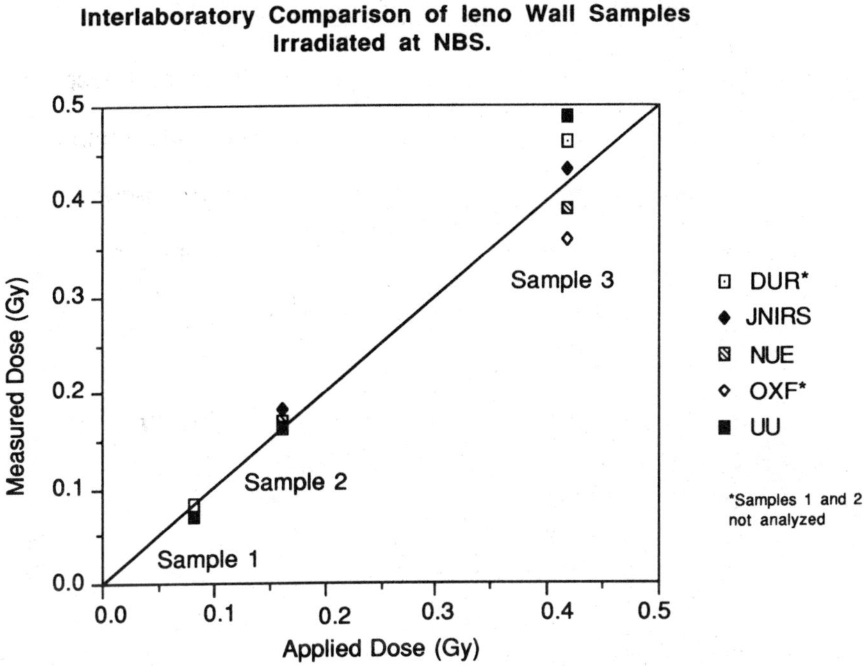

Fig. 14. Interlaboratory comparison of Ieno Wall samples irradiated at NBS

The other intercomparison used bricks from the Ieno Wall in Nagasaki (Figures 16, 17, 18) and the results in this case showed no outliers. The combined laboratory measurements for each of three layers of the brick produced standard deviations of approximately 10%.

Comparison of Measurements with Theory

A few words are necessary to explain some of the factors which go into comparing the TL measurements with the theoretical calculations. First of all, the quantity of interest with the new dosimetry system, DS86, is not dose to bricks and tiles but dose to people and to specific organs in people.

Since a step in the process of arriving at dose to people using DS86 involves calculating free-field kerma to tissue, it was necessary to arrive at a comparison of this quantity using the TL measurements in bricks and tiles. To do this, a great deal of modeling was required. Density, elemental constituents and depth of the sample were input

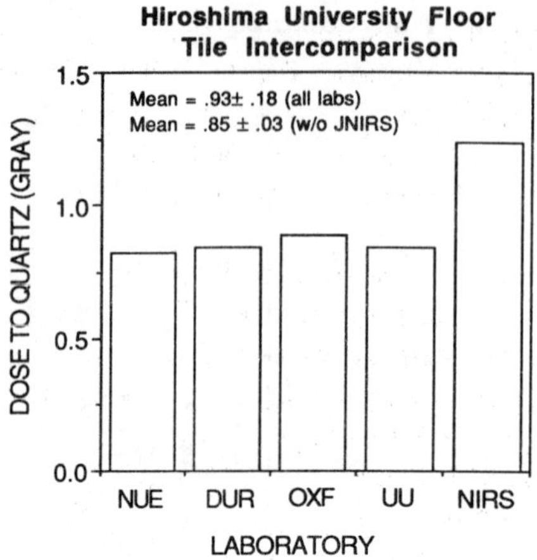

Fig. 15. Hiroshima University floor tile intercomparison

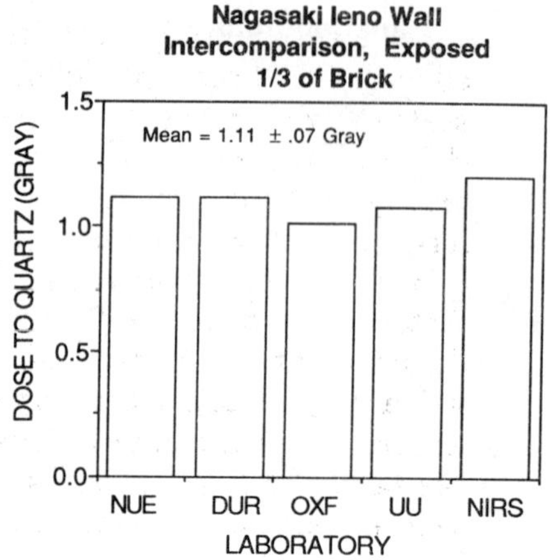

Fig. 16. Nagasaki Ieno Wall intercomparison, exposed 1/3 of brick

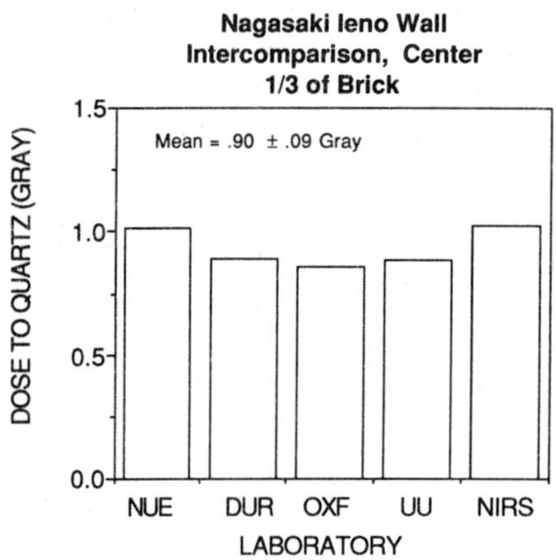

Fig. 17. Nagasaki Ieno Wall intercomparison, center 1/3 of brick

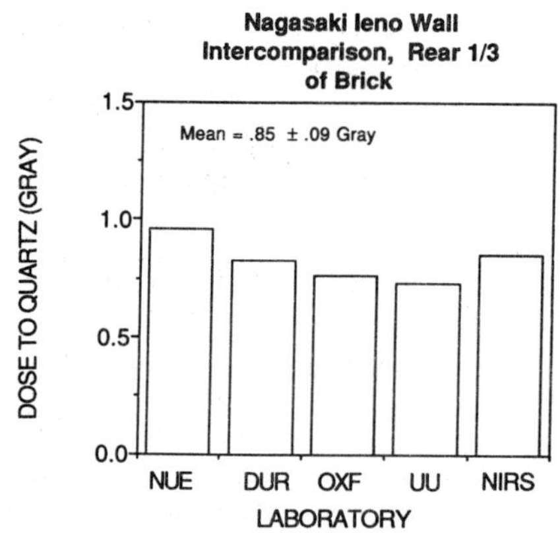

Fig. 18. Nagasaki Ieno Wall intercomparison, rear 1/3 of brick

into the calculations. The geometry, height and composition of the buildings was also modeled.

Once a DS86 calculation of dose to bricks or tiles was arrived at, the ratio of the two was reconverted to free field tissue kerma at the appropriate ground distance from the hypocenter.

For Nagasaki (Figure 19) the values at distances beyond the 1 km support DS86 over the earlier dosimetry system, T65-D. With a linear scale showing ratios of measured to calculated values (Figure 20) it is clear that beyond 1 km agreement of measurement to theory is approximately 10%. The crosses are values of the early measurements performed at NIRS in the 1960's.

At Hiroshima there is again support of DS86 over T65-D values at distances beyond 1 km (Figure 21), although not necessarily at close-in distances. On the linear scale showing ratio of measured to calculated values (Figure 22), a dichotomy between ratios at distances less than 1 km and those at distances greater than 1 km is apparent. Nevertheless the agreement between measurements and theory is good, with calculations exceeding measured values by approximately 10% at distances of less than 1 km, and measurements exceeding calculated values by approximately 18% at distances beyond 1 km.

Because the preponderance of samples beyond 1 km came from a single building at 1.4 km, we feel that it would be premature to suggest changes in the theoretical parameters of DS86 at this time. If the TL results from tiles from a newly sampled building at 1.6 km support the measurements at 1.4 km then such changes could be considered at a future date.

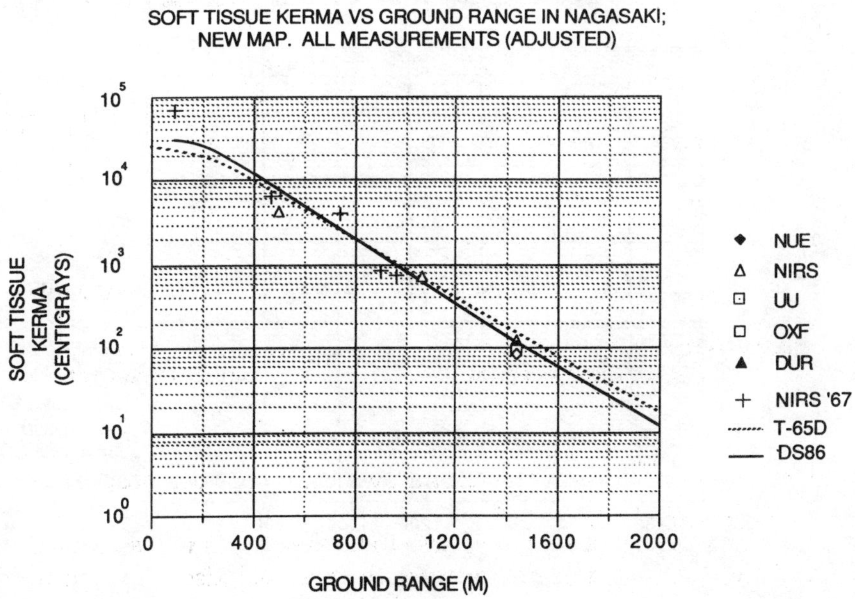

Fig. 19. TL measurements (soft issue kerma) versus ground range in Nagasaki

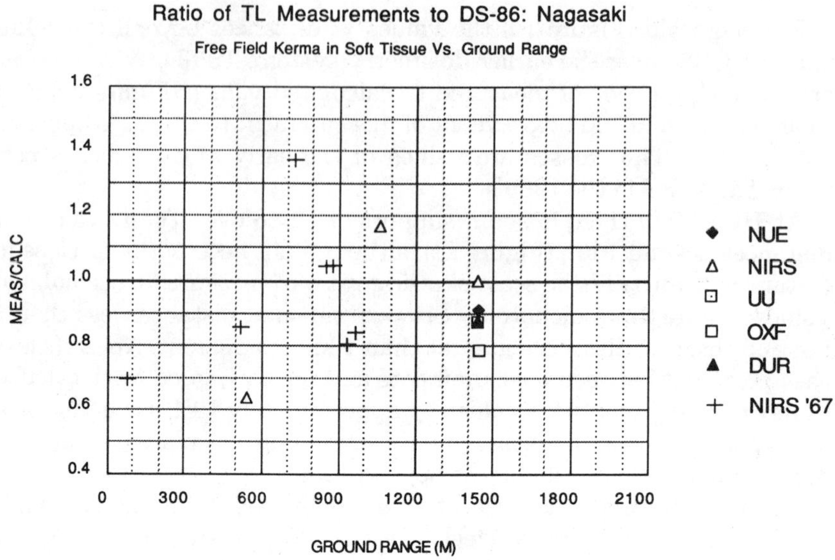

Fig. 20. Ratio of TL measurements to DS86 calculated values for Nagasaki

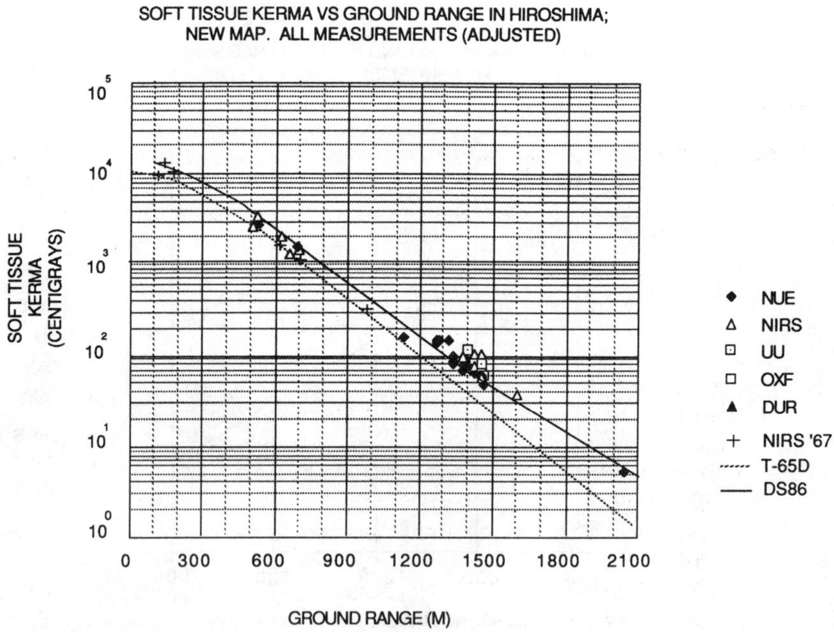

Fig. 21. TL measurements (soft tissue kerma) versus ground range in Hiroshima

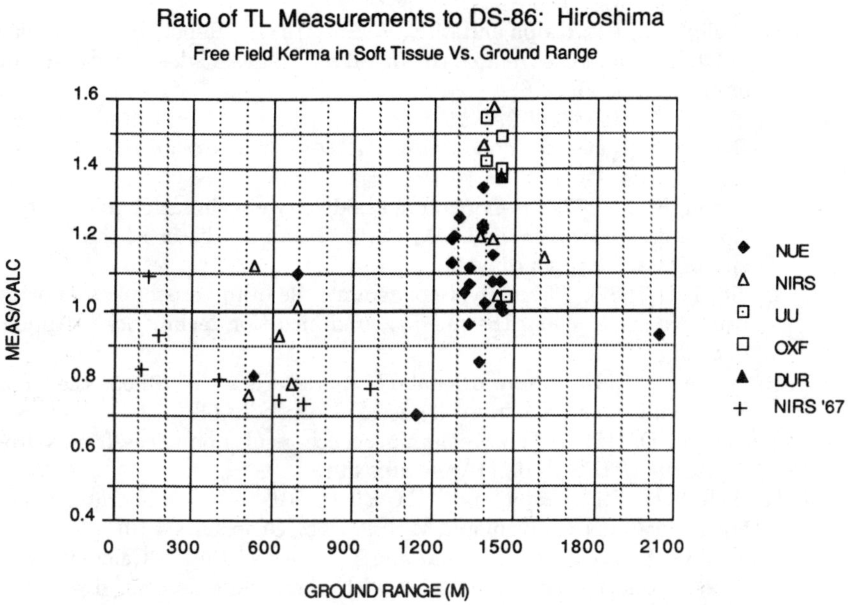

Fig. 22. Ratio of TL measurements to DS86 calculated values for Hiroshima

References

1. Maruyama, T., Y. Kumamoto, Y. Ichikawa, T. Nagatomo, M. Hoshi, E. Haskell and P. Kaipa. (1987). "Thermoluminescence of Gamma Rays." In *The Dosimetry of the Atomic Bombs at Hiroshima and Nagasaki,* Vols. I and II. William C. Roesch, Ed., Radiation Effects Research Foundation.

with the following appendices:

Maruyama, T., Y. Kumamoto and Y. Noda. (1987)."Reassessment of gamma-ray doses from atomic bombs in Hiroshima and Nagasaki." (Appendix 4-1)

Ichikawa Y. and T. Nagatomo. (1987). "Thermoluminescence dosimetry of gamma rays from the atomic bomb in Hiroshima at 1.27 to 1.46 kilometers from the hypocenter." (Appendix 4-2)

Nagatomo T. and Y. Ichikawa. (1987). "Thermoluminescence dosimetry of Hiroshima atomic bomb gamma rays by the predose method." (Appendix 4-4)

Hoshi M., Y. Ichikawa and T. Nagatomo. (1987). "Thermoluminescence dosimetry of the Hiroshima atomic bomb gamma rays about 2 km from the hypocenter." (Appendix 4-5)

Haskell, E.H., P.L. Kaipa and M.E. Wrenn. (1987). "Report on University of Utah TL analyses related to Hiroshima/Nagasaki dose reassessment effort." (Appendix 4-6)

Eagleson, D. (1987). "Report from the Armed Forces Radiobiology Research Institute concerning LINAC and Co-60 irradiations." (Appendix 4-6a)

Hoffman, R.J. (1987). "Report on methodology of calibration and irradiation of samples with the UDM Cs-137 beam irradiator at the University of Utah." (Appendix 4-6b)

Bailiff, I.K. (1987). "TL analyses of ceramic tile from Hiroshima and brick from Nagasaki using the pre-dose and inclusion techniques." (Appendix 4-7)

Huxtable, J. (1987). "Conventional TL characteristics of a floor tile from Hiroshima and a brick from Nagasaki." (Appendix 4-8)

Stoneham, D. (1987). "Thermoluminescence results on slices from Hiroshima tile UHFS FT03." (Appendix 4-9)

Haskell, E.H., P.L. Kaipa, and W.H. Ellett. (1987). "Interlaboratory calibration using NBS-irradiated Mg_2SiO_4:Tb." (Appendix 4-10)

Kaul, D.C., S.D. Egbert, T. Kuhn and J. Roberts. (1987). "Calculation of dose in quartz for comparison with thermoluminescent dosimetry measurements." (Appendix 4-11)

2. Aitken, M.J. (1985). *Thermoluminescence Dating.* London: Academic Press.
3. Hashizume T., T. Maruyama, A. Shiragai, E. Tanaka, M. Izawa, S. Kawamura and S. Nagaoka. (1967). "Estimation of the air dose from the atomic bombs in Hiroshima and Nagasaki." *Health Physics* 13, 149.
4. Ichikawa Y., T. Higashimura, T. Shidei. (1966). "Thermoluminescence Dosimetry of gamma rays from atomic bombs in Hiroshima and Nagasaki." *Health Physics* 12, 395.

Neutron Activation Studies Related to the Reassessment of Hiroshima and Nagasaki Atomic-Bomb Dosimetry

Toshiso Kosako
Research Center for Nuclear Science and Technology
University of Tokyo

1. Introduction

Neutron activation studies [1] related to the reassessment of Hiroshima and Nagasaki atomic-bomb dosimetry were used to check the results of the radiation transport calculation through air. The neutron activation studies have two big categories; one is real field measurements of Hiroshima and Nagasaki and the other is simulation experiments of these atomic-bomb radiation fields.

About the first category, the actual field measurements, we could obtain three different activations. The first one is ① sulfur activation, which was obtained in the very early stage after the blast in 1945. The second one is ② cobalt activation, which was obtained in the 1960's. The third one is ③ europium activation, which was obtained very recently.

The other simulation experiments were also done to clarify the atomic-bomb radiation fields. The first example is ④ activation foil test at Nevada test site in 1950's, and the recent experiment is ⑤ neutron measurements at APRD (Army Pulse Radiation Division, USA) in 1980.

2. Neutron Activation Studies

The neutron activation studies are surveyed according to the typical five categories above-mentioned.

① *Sulfur activation*

The (n,p) reaction of sulfur produces ^{32}P [^{32}S (n,p) ^{32}P]. The reaction cross section is a threshold type and is about 0.25 b above about 3 MeV. The produced radioisotopes of ^{32}P have a half-life of 14.28 days and emit 1.71 MeV β-ray. This half-life is so short that we can use only a very restricted data measured in 1945. Sulfur samples were gathered from some electric insulators lying in the burned field. The beta activity were measured employing a classical electrostatic technique by a Lauritsen electroscope (see Figure 1). The main data we can use were obtained by Yamasaki and Sugimoto [2] and were revised by Hamada [3,4], which

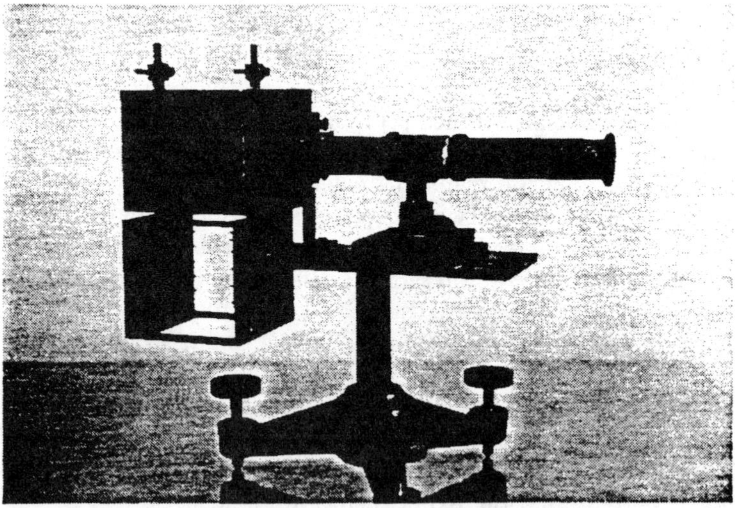

Fig. 1. The Lauritsen Electroscope used by Yamasaki and Sugimoto (Hamada; [3]).

shown in Table 1. Other data were also given by Arakatsu. The characters of this reaction are as follows: The good point of these data is the fast neutron reactions at above 3 MeV region, which are an important part of neutron kerma. But there are some weak points; the vagueness of sampling circumstances, large error bars resulted by an old measuring technique, the self-shielding effect of insulator and the angular dependence of activation to incident neutrons, and some impurities in the sulfur.

② *Cobalt activation*

The considering reaction is ^{59}Co (n,γ) ^{60}Co and its cross section is about 37 b. The main contributing neutrons are thermal to epi-thermal neutrons. The produced ^{60}Co has a half-life of 5.271 years, and they emit β-ray of 0.318 MeV and γ-ray of 1.173 and 1.332 MeV. The main data were obtained by Hashizume et al. [5,6]. The steel bars in the concrete were gathered in Hiroshima and Nagasaki. These bars were solved into hydrochloric acid to extract cobalt element. The radioactivity of this chemically enriched cobalt sample was measured by a β-ray plastic scintillation spectrometer shown in Figure 2. The measured results are listed in Table 2 [5]. The characters of cobalt activity data are as follows. These data covers a long distance range ($\sim$1 km) with little error bars (within $\sim$18%). On the other hand, the weak points are as

TABLE 1—*Revised Data for ^{32}P Activity in Sulfur (Hamada [4])*

Sample	Electrometer Reading div/s x 10-3		32P Act.* Bq/gs	Decay Corr. 32P Act. dpm/gs	Ground Distance in meters
	gross	net			
A-4	2.00	0.76	6.7	2200	305
B-5	2.24	1.00	8.8	2940	178
C-6	1.54	0.30	2.6	880	358
D-12	1.62	0.38	3.3	1140	433
"	1.78	0.54	4.7	1500	"
E-13	2.05	0.81	7.1	2430	76
F-14	1.66	0.42	3.7	1260	417
G-15	1.67	0.45	3.9	1370	(440)
"	1.75	0.53	4.6	1620	(")
H-7	1.44	0.20	1.8	630	721
J-8	1.30	0.06	0.53	190	1025
K-29	1.36	0.12	1.62	560	682

*Estimated standard deviation would be no more than $\pm$1Bq.

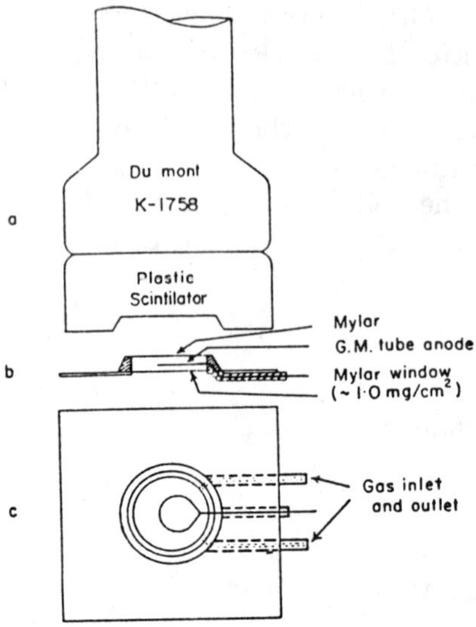

Coincidence type β-ray spectrometer. (a) and (b) are side views of the detector. [The GM tube (b) should be inserted in the hollow cut of the scintillator of (a).] (c) is a top view of the GM tube (b).

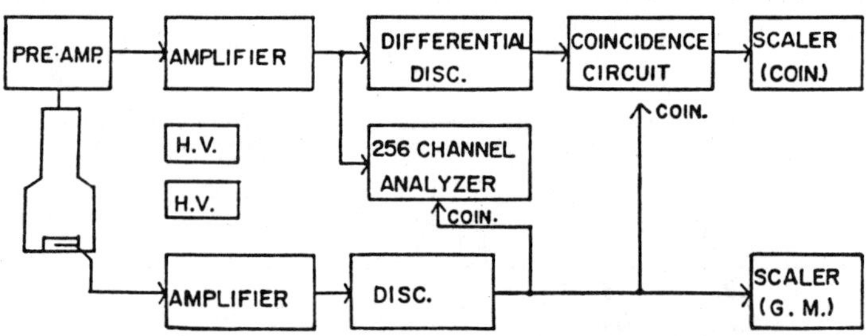

Block-diagram of electronic circuit.

Fig. 2. Beta-ray Plastic Scintillator System and its Electric Circuits Used by Hashizume [5].

TABLE 2— ^{60}Co *Activity per Milligram Co in Hiroshima and Nagasaki as a Function of Ground Distance (Hashizume [5]).*

(Hiroshima 132°27.43′ x 34°23.47′ Nagasaki 129°50.95′ x 32°46.14′). Where the height of epicenter assumed 590 m at Hiroshima and 500 m at Nagasaki

City	Coordinates of iron samples	Distance from hypo-center (m)	Incident angle	Activity (cpm/mg.Co)	Q_θ (rate)	N_θ (R per cpm/ mg.Co)	Neutron dose (rad)
Hiroshima	27.61 x 23.47	260	66.60	2.09 ± 0.170	0.463	630	7864
	27.54 x 23.81	640	44.54	0.324 ± 0.0204	0.485	586	1188
	27.30 x 23.89	779	41.69	0.146 ± 0.0013	0.495	580	541
	27.50 x 46.12	1180	27.65	0.0124 ± 0.0023	0.566	562	51
Nagasaki	51.29 x 46.02	590	42.55	0.140 ± 0.0066	0.480	593	514
	50.42 x 46.49	1030	31.79	0.0127 ± 0.0010	0.535	568	52

follows; competing reactions in the steel material such as $^{63}Cu\ (n,\alpha)\ ^{60}Co$ ($n > 5$ MeV, cross section ~15 mb) and $^{60}Ni(n,p)\ ^{60}Co$ ($n > 1$ MeV, cross section ~0.5b), and the vagueness of concrete composition (that was not measured at that time and the sample doesn't exist now).

③ *Europium activation*

The following two reactions can be observed even now. One is $^{151}Eu\ (n,\gamma)\ ^{152}Eu$ reaction and this has a cross section of about 5,900 b for thermal neutrons. The produced ^{152}Eu has a half-life of 13.2 years and emits β-ray of 0.696 MeV and γ-ray of 0.122, 0.344 MeV and so on. The second reaction is $^{153}Eu\ (n,\gamma)\ ^{154}Eu$ and its cross section is about 2,000 b for thermal neutrons. The produced ^{154}Eu has a half-life of 8.5 years and emits β-ray of 0.577 MeV and γ-ray of 0.123, 0.723 MeV and so on. The data of europium in Hiroshima and Nagasaki have been recently obtained by several groups; Kanazawa University group (Sakanoue, Nakanishi et al.) [7], Nagasaki University group (Okajima and Miyajima) [8] and Hiroshima University group (Hoshi et al.). Most of the europium samples were detected in the rock and the measurements were performed by using intrinsic or lithium-drift germanium γ-ray spectrometers (See Figure 3). The strong point of these data lies in the clearance of their sampling locations and circumstances. But these activities are very low and it takes a lot of time to get adequate statistical counts. Therefore far distance data more than ~1 km are quite difficult to measure with a good accuracy in spite of using chemical enrichment techniques. In Figure 4, the new data of spatial distribution of europium activity in Hiroshima measured by Nakanishi are presented with the comparison of calculation.

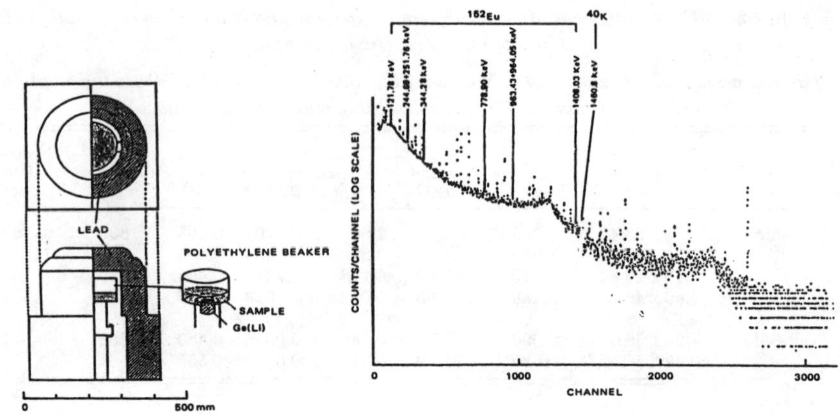

Shielded Ge(Li) gamma ray spectrometer and sample arrangement

Gamma spectra of rock (36 m from the hypocenter) measured by Ge(Li) 60 cc detector, Nagasaki

Fig. 3. Ge(Li) Detector System and Typical Gamma-Ray Spectrum of ^{152}Eu in Rock at Nagasaki (Okajima [8]).

④ *Other simulation experiment (activation tests in the Nevada test site)*

Several nuclear weapons tests in the atmosphere were done in the Nevada test site in 1950's. The neutron activation tests were done in these experiments. Some of these data can be used as benchmark data for the dosimetry works of Hiroshima and Nagasaki. Dolatshahi and Kaul referred [9] to the data of the Shot Fox of the Ranger Series (22 kt, 437 m height of epicenter, 85% humidity) and the Shot Dog of the Buster-Jangle Series (21 kt, 432 m height of epicenter, 43% humidity) as a Nagasaki atomic bomb simulation. They got several neutron activation data, i.e. sulfur and gold. Considering the measuring techniques at that time, β-ray counting must be major methods. Kaul utilized gold activation data of the above two tests [9] to verify the transport calculations (see Figure 5). The considering reaction is ^{197}Au (n,γ) ^{198}Au and its cross section is 96 b for thermal neutrons. The produced ^{198}Au has a half-life of 2.696 days and emits β ray of 0.961 MeV and γ rays of 0.412 MeV and so on.

The characters of these data are as follows. The most strong point is that these are the real size benchmark tests, especially for Nagasaki's bomb, and the experimental conditions are extrememly clear. But almost all these data are not published, there exist some difficulties to discuss the details. We must do some corrections to the obtained data because of the different conditions between the Nevada test site and Hiroshima or Nagasaki (i.e., atmosphere, ground covers, a blast height, etc.).

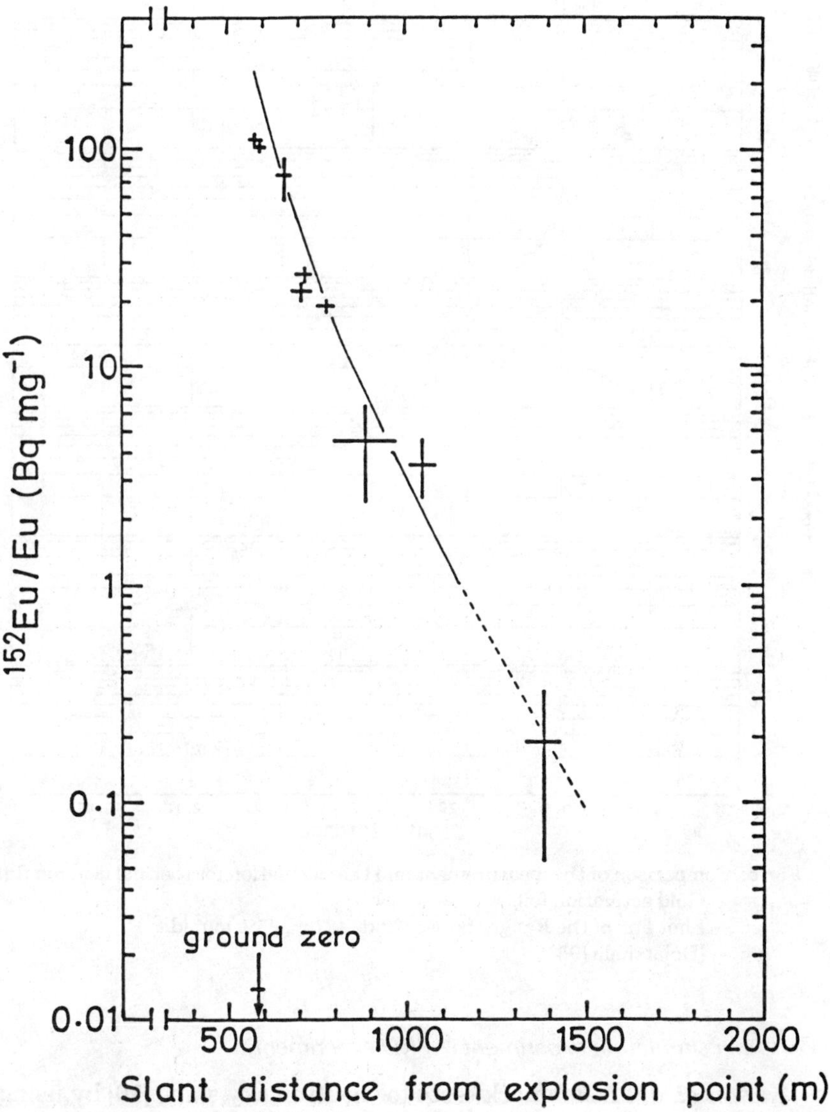

Fig. 4. Specific Radioactivity of ^{152}Eu at Hiroshima along a Slant Distance (Nakanishi [7]).

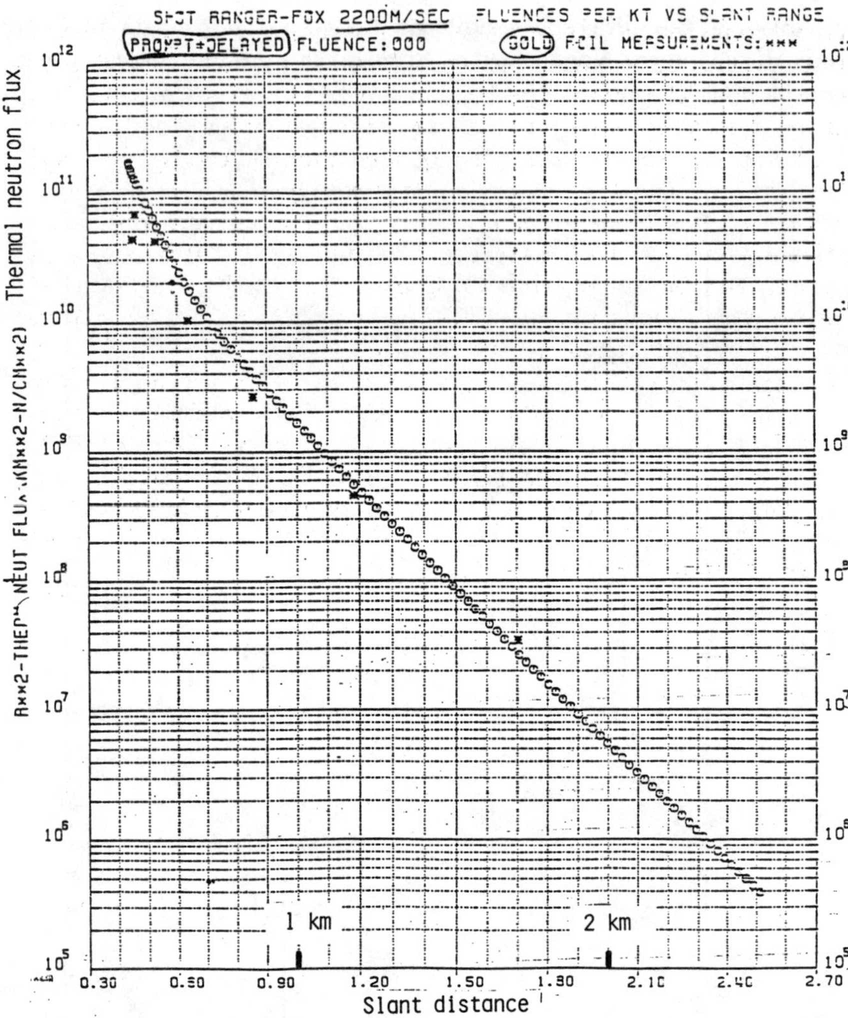

Fig. 5. Comparison of the measurement and the calculation for thermal neutron flux
— Gold activation foil,
— Shot Fox of the Ranger Series (22kt, 437 m, 85% humidity)
— (Dolatshahi [9])

⑤ *Other simulation experiment (APRD experiment)*

Kazi *et al.* did benchmark tests for neutron sky shine [10] by using a GODIVA type bare reactor at APRD (Army Pulse Radiation Division, USA), which can lift to a 14 m high position. The measurements were done along a ground surface by various dosimeters and spectrometers for several distances within 1.6 km. The results were summarized to tissue kerma. The most significant meaning of this experiment is a neutron spectral information, that was obtained by an NE-213 liquid scintillator, a proton recoil proportional counter, Bonner sphere spectrometers and so on. For the result of this experiment, the intercomparisons of calculation were done between LLNL (Lawrence Livermore National Lab.), ORNL (Oak Ridge National Lab.), DREO (Defense Research Establishment, Ottawa, Canada) (for DOT code) and SAI (Science Applications Inc.) (for DOT code and Monte Carlo code). In Figure 6, the result of 400 m position is shown.

We can also use other measured results of neutron activations, neutron dosimetries and neutron spectrometries for the validation of calculations [i.e. the results of Ichiban-project, some experiments of HPRR (Health Physics Research Reactor) at ORNL, some sky-shine experiment data, some data of Whalen's benchmark field of little boy atomic bomb (Hiroshima bomb) head [11,12]. In any case, we must pay attention to the accuracy and the reliability of the data.

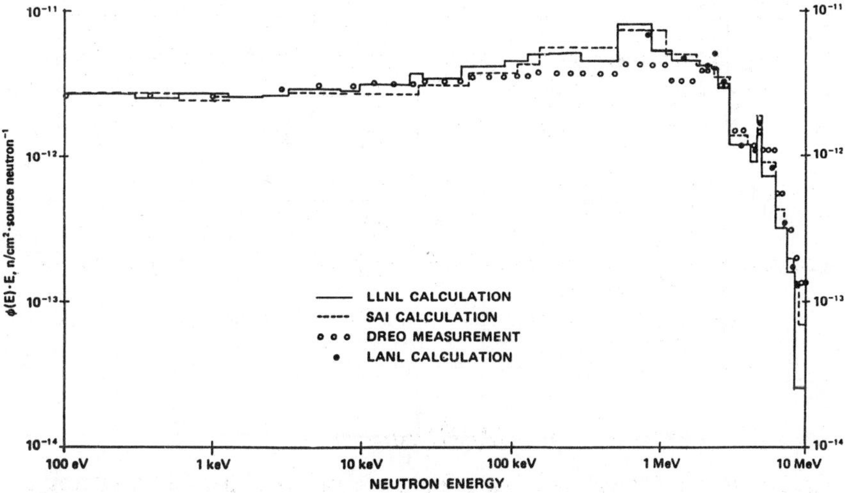

Fig. 6. Neutron Sky-Shine Experiment at APRD by Kazi [10]. The measured and calculated spectra at 400 m ground range are compared.

3. Comparison between the Measurements of Neutron Activation and Neutron Transport Calculation through Air.

The neutron transport calculations in air were done by LLNL, ORNL and SAI. The computer codes used were DOT4 (discrete ordinate method), TART (Monte Carlo method), and so on. The details of these calculations are described in each report [i.e. Ref.13]. Here, a generous situation of the comparison of measured data with calculation is described by using some figures.

Figure 7 shows the data of ^{32}P per gram of sulfur at Hiroshima. The Hamada's new data [4] are plotted with SAI calculation. The discrepancy of measurement and calculation is about 50% at 500 m ground distance and about a factor of 5 at 1 km. We can't discuss a detail from these data because the error bars of ^{32}P are big.

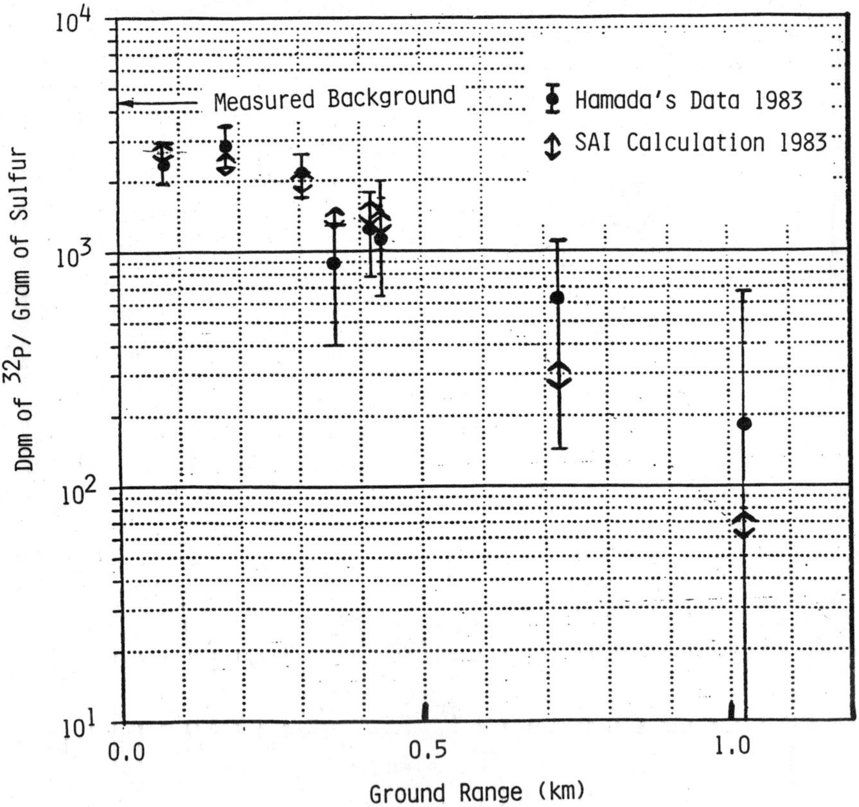

Fig. 7. Comparison of Measured and Calculated Sulfur Activities

Figure 8 shows the ratio of calculation and measurement of ^{60}Co activities. Both data of Hiroshima and Nagasaki show a similar tendency; an overestimation at the near range and an underestimation at a far distance. The discrepancy is great and is a factor of 5 at 1 km, but a clear explanation has not been made.

Figures 9 and 10 show the data of ^{152}Eu per milligram of europium. These experimental data at both Hiroshima and Nagasaki spread in the range of several tens of per cent, centering around the LLNL calculated line.

Altogether, the accuracy of the calculations used here will be about 10-20% for neutron tissue kerma in air excluding the uncertainty of atomic bomb source intensity.

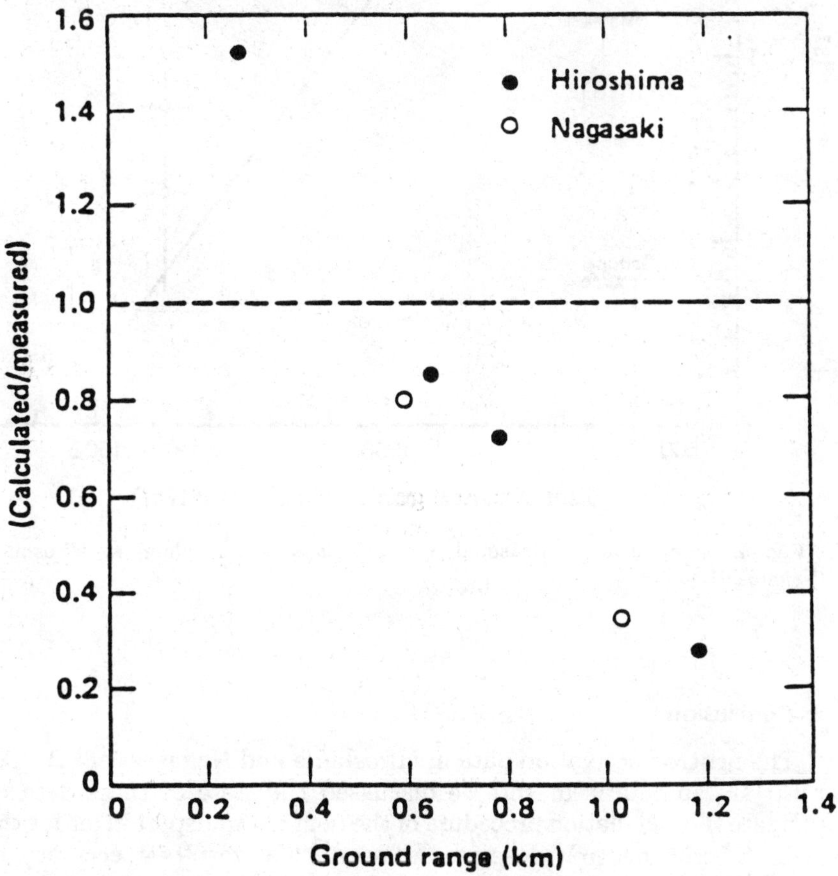

Fig. 8. Comparison of Measured and Calculated Cobalt-60 Activities (Loewe [14]).

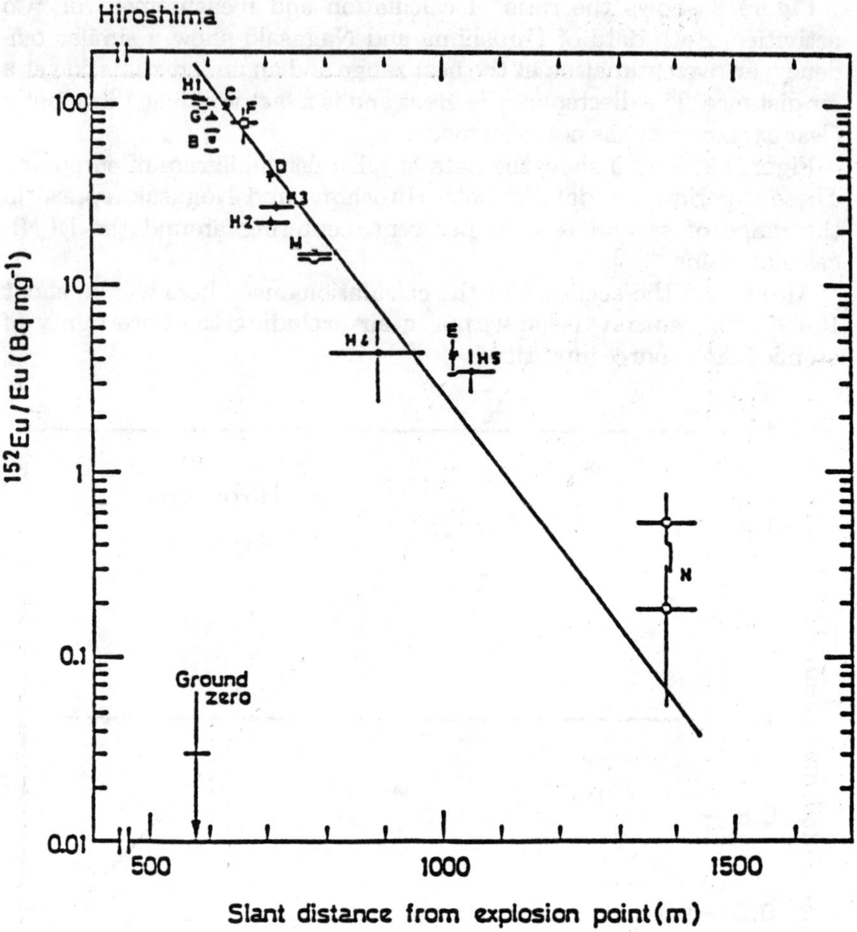

Fig. 9. Comparison of Measured and Calculated ^{152}Eu/Eu(mg) at Hiroshima (Nakanishi [7]).

4. Conclusion

The neutron activation data in Hiroshima and Nagasaki of ^{32}P, ^{60}Co, ^{152}Eu are summarized, and we discussed the state of these data to validate the calculation procedure of the neutron transport in air for the atomic bomb neutron dose reassessment. The resultant accuracy of obtained calculations (DS86 system) is 10-20% for tissue kerma in air at Hiroshima and Nagasaki.

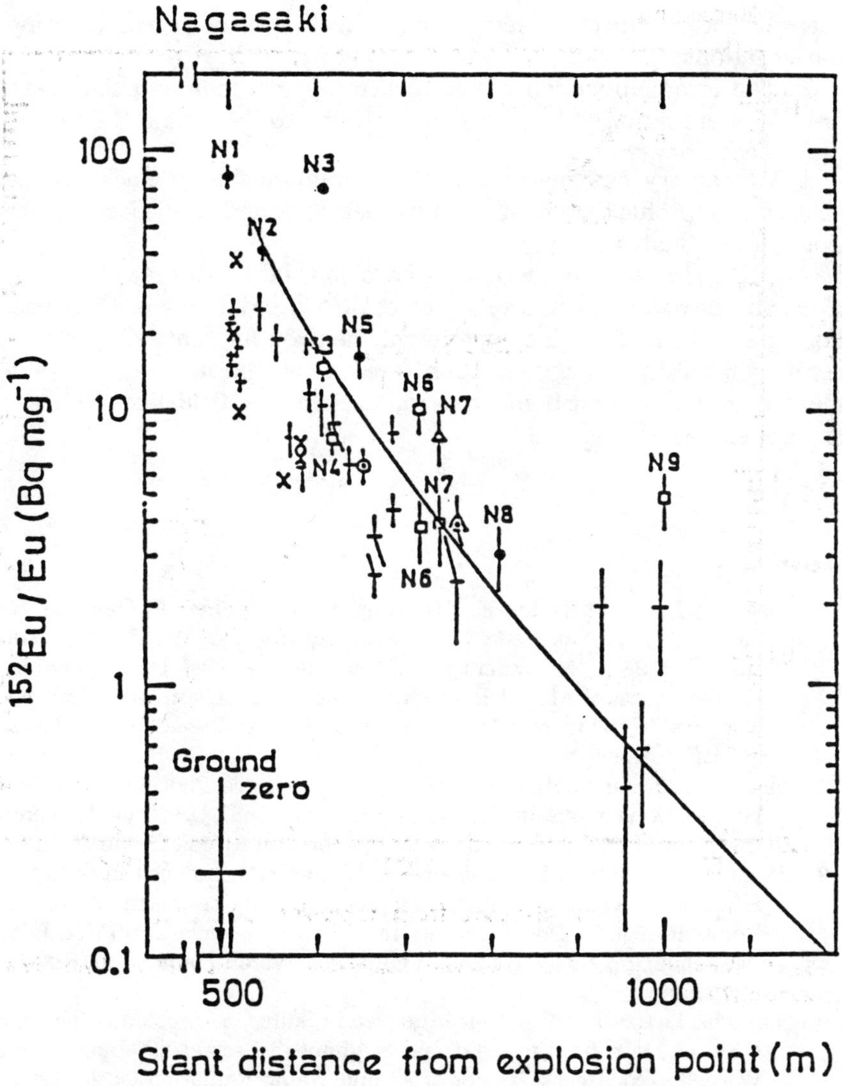

Fig. 10. Comparison of Measured and Calculated ^{152}Eu/Eu(mg) at Nagasaki (Nakanishi [7]).

If we have a chance to extend this reassessment work to a future, the following points are recommendable for future works.

① First of all, we must do our best for collecting the irradiated samples and making those documentation, because the time flow of 40 years will wash out everything (sample itself and the people knowing the situations).

② The intercomparison of the measuring methods and the measured data are required for europium activity to determine the experimental accuracy.

③ We can try new methods such as uranium fission tracks in the glass, depth profile data of europium in the rock and stable isotope (for example ^{14}C) analysis.

④ Other simulation experiments are also valid for this reassessment to clarify the detailed situations of calculations and used cross sections. For examples, a sky-shine experiment, neutron air scattering experiment using a liquid air simulation tank, a neutron activation experiment using an atomic-bomb neutron benchmark field like Whalen's experiment [11,12].

References

1. Loewe, W.E., Mendelsohn, E., Hamada, T., Maruyama, T., Okajima, S., Pace, III, J.V., Sakanoue, M., Kondo, S., Woolson, W. A. and Hashizume, T. (1987). "Measurements of Neutron Fluences," US-Japan Joint Reassessment of Atomic Bomb Radiation Dosimetry in Hiroshima and Nagasaki (Final report), Vol. 1, Radiation Effects Research Foundation (RERF), Hiroshima.
2. Yamasaki, F. and Sugimoto A. (1945). "Radioactive Phosphorus P-32 in Human Bone and Sulfur Insulators in Hiroshima," US Naval Research Institute, NMRI Series of Japanese Reports on Atomic Bomb, No. 17.
3. Hamada, T. (1983). "Measurements of ^{32}P Activity Induced in Sulfur in Hiroshima," US-Japan Joint Workshop for Reassessment of Atomic Bomb Radiation Dosimetry in Hiroshima and Nagasaki (First Joint Workshop), pp. 45-56, Radiation Effects Research Foundation (RERF), Hiroshima.
4. Hamada, T. (1983). "^{32}P Activity Induced in Sulfur in Hiroshima: Reevaluation of Data by Yamasaki and Sugimoto," Second US-Japan Joint Workshop for Reassessment of Atomic Bomb Radiation Dosimetry in Hiroshima and Nagasaki (Second Joint Workshop), pp. 52-55, Radiation Effects Research Foundation (RERF), Hiroshima.
5. Hashizume, T., Maruyama, T., Shiragai, A., Izawa M., Kawamura, S. and Nagaoka, S. (1967). "Estimation of Air Dose from the Atomic Bombs in Hiroshima and Nagasaki," Health Physics, 13, 149-161.
6. Hashizume, T. (1983). "Present Plans for Dose Reassessment Experiments by the Japanese," US-Japan Joint Workshop for Atomic Bomb Radia-

tion Dosimetry in Hiroshima and Nagasaki (First Joint Workshop), pp. 7-12, Radiation Effects Research Foundation (RERF), Hiroshima.

7. Nakanishi, T., Imura, T., Komura, K. and Sakanoue, M. (1983). "^{152}Eu in Samples Exposed to Nuclear Explosions at Hiroshima and Nagasaki," Nature, 302, 132-134.
8. Okajima, S. and Miyajima, J. (1983). "Measurement of Neutron-Induced ^{152}Eu Radioactivity in Nagasaki," US-Japan Joint Workshop for Atomic Bomb Radiation Dosimetry in Hiroshima and Nagasaki (First Joint Workshop), pp. 156-167, Radiation Effects Research Foundation (RERF), Hiroshima.
9. Dolatshahi, F., Kaul, D.C. and Egbert, S.D. (1987). "Delayed Neutrons," US-Japan Joint Reassessment of Atomic Bomb Radiation Dosimetry in Hiroshima and Nagasaki (Final report Vol. 2), pp. 95-112, Radiation Effects Research Foundation (RERF), Hiroshima.
10. Kazi, A.H., Heimbach, C.R., Harrison, R.C. and Robitaille, H.A. (1985). "Comparison of Measured and Calculated Radiation Transport in Air-Over-Ground Geometry to 1.6 km from a Fission Source," Nuclear Science and Engineering, 85, 371-386.
11. Whalen, P.P. (1983). "Source Terms for the Initial Radiations," US-Japan Joint Workshop for Atomic Bomb Radiation Dosimetry in Hiroshima and Nagasaki (First Joint Workshop), pp. 13-39, Radiation Effects Research Foundation (RERF), Hiroshima.
12. Whalen, P.P., Soran, P.D., Malenfant, R. and Forehand, Jr., H.M. (1983). "Experiments at Los Alamos National Laboratory with the Replica of the Hiroshima Weapon," Second US-Japan Joint Workshop for Reassessment of Atomic Bomb Radiation Dosimetry in Hiroshima and Nagasaki (Second Joint Workshop), pp. 21-25, Radiation Effects Research Foundation (RERF), Hiroshima.
13. Kerr, G.D., Pace, III, J.V., Mendelsohn E., Loewe, W.E., Kaul, D.C., Dolatsuhahi, F., Egbert, S.D., Gritzner, M., Scott, Jr., W.H., Marcum, J., Kosako, K. and Kanda, K. (1987). "Transport of Initial Radiations in Air over Ground," US-Japan Joint Reassessment of Atomic Bomb Radiation Dosimetry in Hiroshima and Nagasaki (Final Report, Vol. 1), pp. 66-142, Radiation Effects Research Foundation (RERF), Hiroshima.
14. Loewe, W.E. (1985). "Hiroshima and Nagasaki Initial Radiations: Delayed Neutron Contributions and Comparison of Calculated and Measured Cobalt Activities," Nuclear Technology, 68, 311-318.

An Assessment of Dosimetry System 1986 (DS86) Components

Dean C. Kaul
Science Applications International Corporation
Schaumberg, Illinois

It is said that success has many fathers, while failure is an orphan. If that is true then the Dosimetry System 1986 (DS86) (Roesch, 1987) cannot help but be a success because it has many fathers indeed. I feel honored to have been chosen from among all those who claim paternity to provide an assessment of DS86 and its components as they currently exist and a prediction of what lies in the future. I would like to acknowledge the work of laboratories and individuals from the United States, Japan and the United Kingdom whose contributions have created DS86.

T-65D (Milton and Shohoji, 1968), the dosimetry system that DS86 supercedes, is a tour de force of empirical modeling. Its credibility hangs on the degree of correlation which can be claimed to exist between the true events and their chosen empirical analogues. The effort to replace T-65D grew out of a series of challenges to the premise for such correlation, beginning with the calculation of the radiation leakage of the Hiroshima and Nagasaki weapons by Preeg (1976).

DS86 is based on calculations rather than measurements of all quantities of interest, and there is no doubt that it is different from T-65D, as illustrated in Figures 1-4. Those figures depict the free field kerma and mean or representative kerma within a house and in the marrow of an adult survivor as determined by DS86 and T-65D. The DS86 neutron doses are reduced substantially below those of T-65D, particularly at Hiroshima. The relationships between neutron free field, house

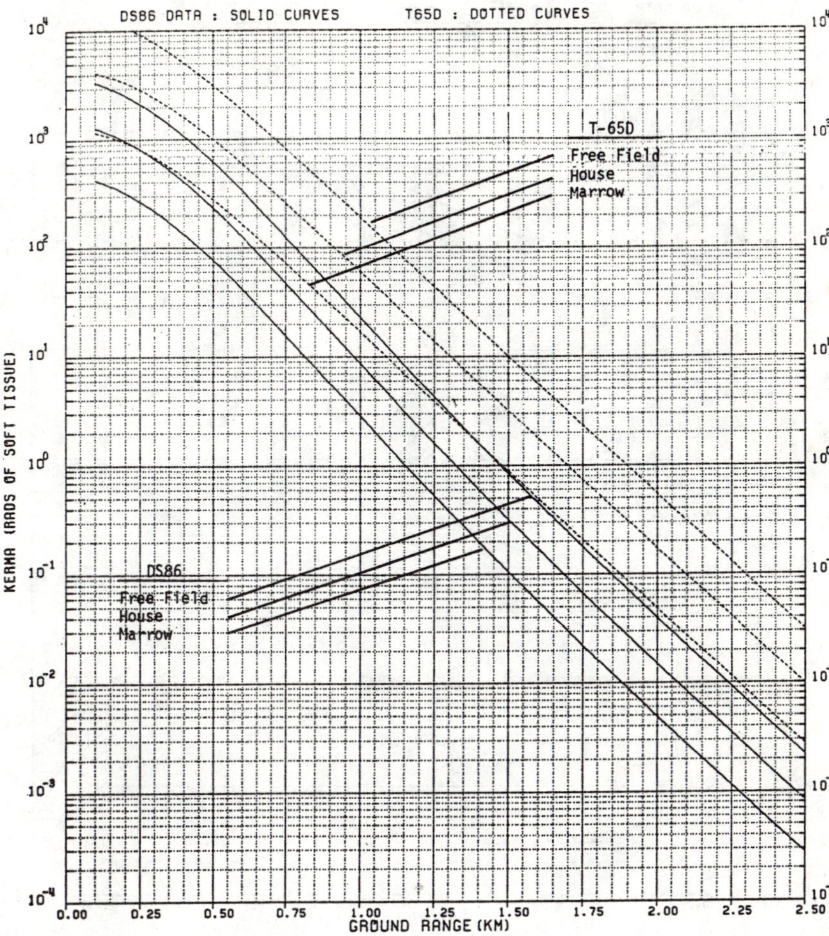

Fig. 1. Hiroshima Neutron Kerma vs. Ground Range Free Field, In House and Marrow Kerma Comparisons

and marrow kerma are similar in the two systems, however. The gamma-ray free fields from the two systems are similar at Nagasaki but considerably different at Hiroshima. Further, DS86 house transmission of gamma radiation is much reduced below that of T-65D, while DS86 transmission into the marrow is much increased. Given these large differences between the two systems, it can reasonably be asked that the calculational methodology on which DS86 is based be capable of reproducing relevant empirical data, particularly that which forms the basis for T-65D. I would like to give you a sum-

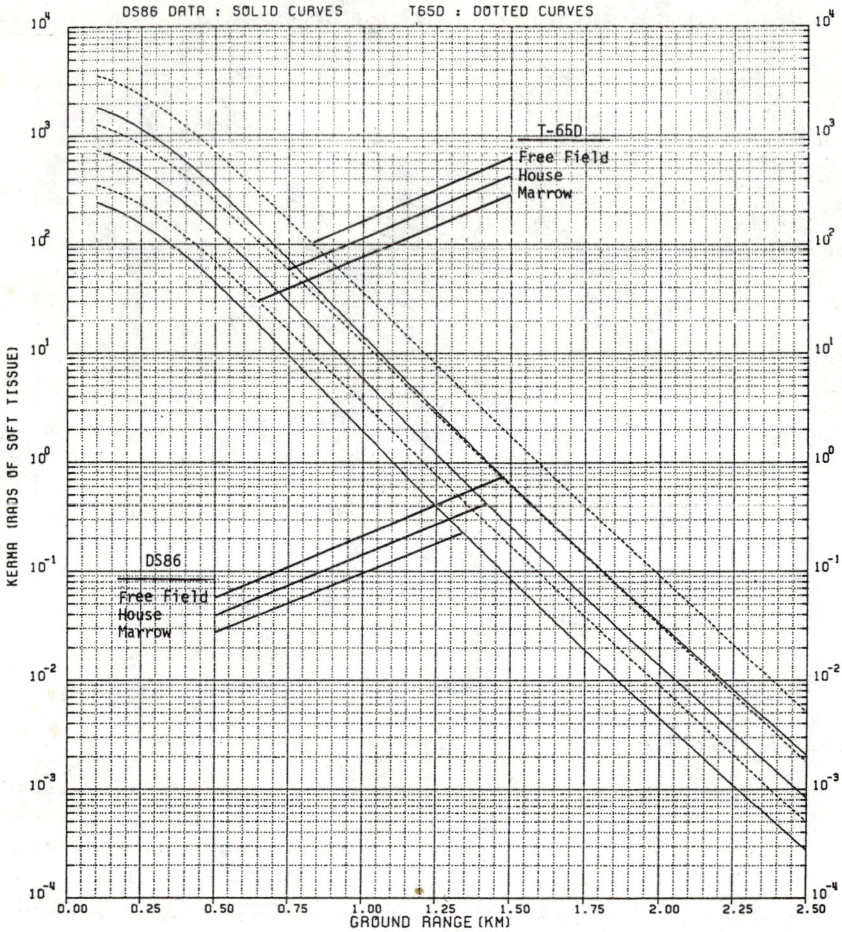

Fig. 2. Nagasaki Neutron Kerma vs. Ground Range Free Field, In House and Marrow Kerma Comparisons

mary of our notable successes as well as our notable failures in that regard and to describe ongoing work which is intended to resolve outstanding problems.

Calculational models have been used to create the basis for DS86 in four separate areas of application. Those are listed in Table 1, along with the important experimental analogues which correspond to each.

TABLE 1—*DS86 Component Models and Their Empirical Analogues*

DS86 Model	
Weapon Yield and Radiation Source Definition	Comet Assembly Experiment Canister Blast Wave Measurement (H)* I Beam Blast Measurement (H)
Free Field Fluence	Thermoluminescent Dosimetry Meas. (H&N) Sulfur Activation Measurement (H) Cobalt Activation Measurement (H&N) Europium Activation Measurement (H&N) Nevada Weapon Test Experiments
Structural Shielding	BREN Experiment Thermoluminescent Dosimetry Meas. (N) Nevada Weapon Test Experiments
Organ Dosimetry	Laboratory Experiments

*Data available from Hiroshima (H) or Nagasaki (N).

Source Definition

The prompt radiation leakage from the two weapons has been calculated using Monte Carlo radiation transport methods and a dynamic geometry model which depicts the disassembly of the weapons as they explode. The Comet assembly experiment, a critical assembly which replicates the Hiroshima weapon, was designed to test the ability of transport models to calculate neutron leakage through large thicknesses of iron and to calculate the approach to a first critical configuration. Comparison with measurements made at the Comet Assembly, as pictured in Figure 5 has confirmed the ability to calculate leakage from the cold assembly. Leakage from the side of the assembly (90°) has calculations and measurements of fluence integrals above E_{min} in agreement within approximately ±10%. Calculated leakage at other angles is low. However, the total leakage from the device is dominated by leakage from the side.

Comparisons with experimental results also confirm the calculated yield range of 13 to 16 kt, the exact value of which depends on the weapon configuration at initialization. Analysis of the deformation of structural steel at Hiroshima and of a pressure history measurement

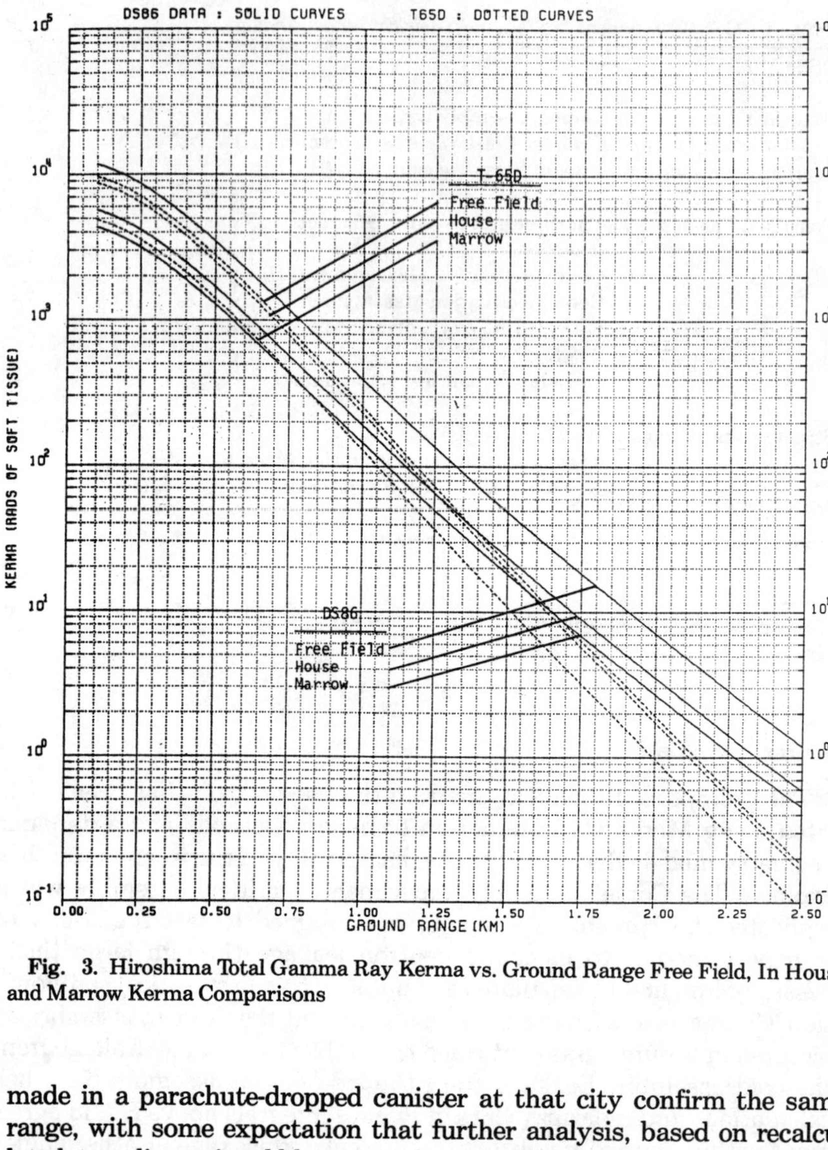

Fig. 3. Hiroshima Total Gamma Ray Kerma vs. Ground Range Free Field, In House and Marrow Kerma Comparisons

made in a parachute-dropped canister at that city confirm the same range, with some expectation that further analysis, based on recalculated, two-dimensional blast wave forms, will converge on an intermediate value. The stated value in DS86 documentation is:

15 kt

Assessments of yield for devices similar to the Nagasaki weapon resulted in an assigned yield of:

21 kt

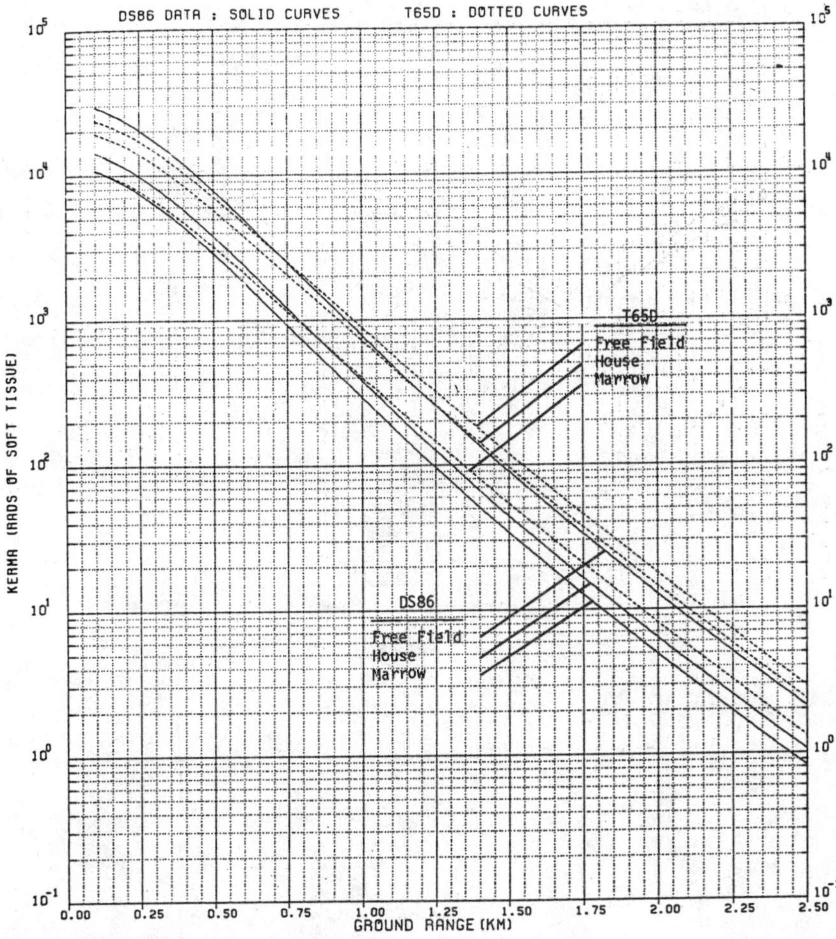

Fig. 4. Nagasaki Total Gamma Ray Kerma vs. Ground Range Free Field, In House and Marrow Kerma Comparisons

No mock-up of the Nagasaki device was ever contemplated because of the large body of available test data pertaining to similar devices.

Free Field

The free field radiation fluence near the ground has been calculated using one- and two-dimensional, discrete ordinate transport methods and multigroup cross sections. Dose and activation measurements

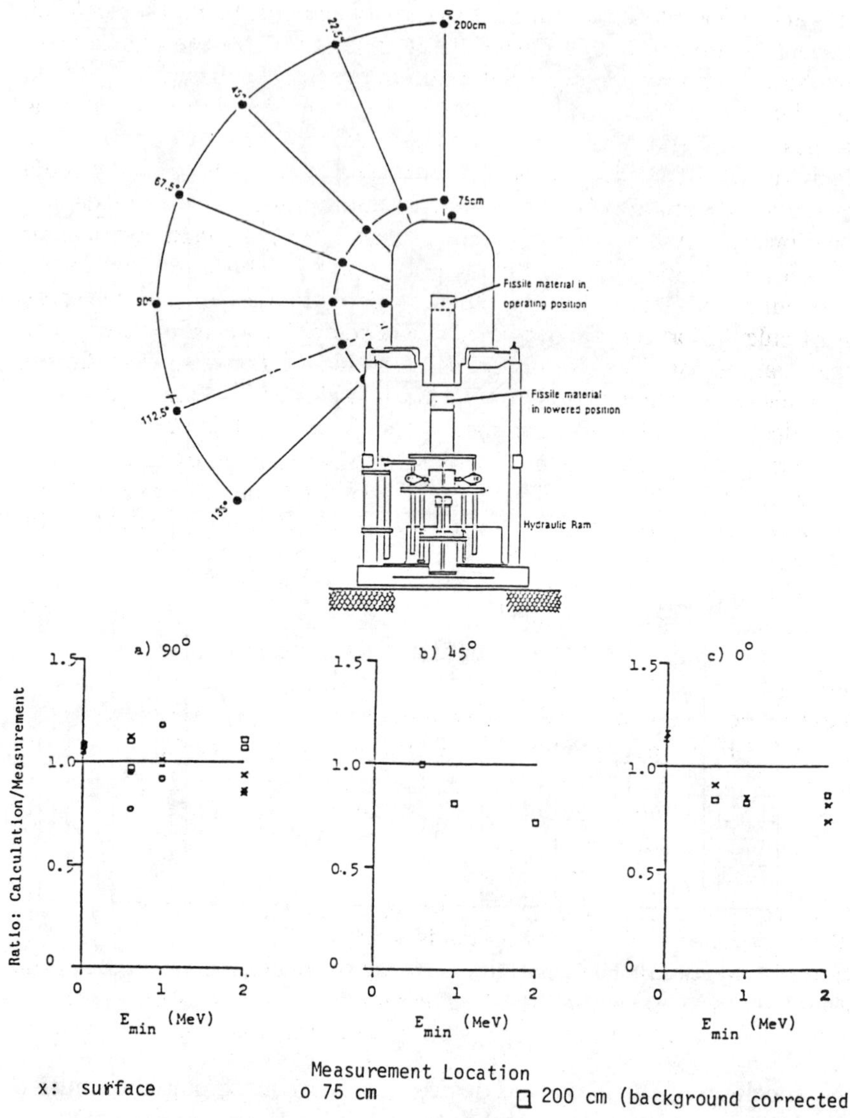

Fig. 5. Hiroshima Weapon Replica with Measurement Positions; Calculation/Measurement ratios of Fluence Integrals ($E_{min} < D < 20$ MeV) at Three Angles

made at the two Japanese cities allow a direct assessment of the quality of DS86 free field data. The most extensive measurements are those of gamma-ray dose in quartz crystals contained in decorative tiles and brick. Dose measurements have been made using thermoluminescent dosimetry techniques. Calculations of the dose to quartz crystals have been made using DS86 gamma ray fluences and Monte Carlo calculations of the shielding provided by local structural detail. The results of the measurements and DS86 calculations in terms of free field equivalent kerma are provided for Nagasaki in Figure 6. T–65D kerma values are also shown. Extensive measurements were made at a location approximately 1,435 meters from the Nagasaki hypocenter known as the Ieno Wall, as shown in Figure 7. Those measurements lend substantial credence to DS86, which provides overall measurement/calculation agreement of 10%. Calculated and measured quartz dose values for Hiroshima, again in terms of the free field kerma equivalent are shown in Figure 8. Both DS86 and T–65D values are provided. More TLD measurements were made at Hiroshima than at Nagasaki and there is more scatter in the data. Measurements at the one location common to all measurers, a spot on the roof of Hiroshima University, approximately 1,450 m from the hypocenter, differ in relationship to the calculations by 10 to 50%, with the calculations being low. However, the majority of the measurements at all locations beyond 1 km agree with the calculations within 10%. The discrepancies between results of the various laboratories, relative to the calculations, are still under study. Measurements are being performed for a new location common to all laboratories approximately 1,500 meters from the Hiroshima hypocenter. In the meantime the existing measurements strongly support DS86 over T–65D gamma-ray doses.

System neutron fluences have been used to calculate sulfur, europium and cobalt activation as measured at the two Japanese cities (no sulfur measurements at Nagasaki). Neutron source energies of 2 MeV and greater are important to producing sulfur activation. The spatial distribution of such fast neutrons is sensitive to the angular distribution of the source. The Hiroshima weapon was more massive at the nose and tail than at the waist. Therefore, the source is biased to the side of the weapon. The result is a nonuniform distribution of sulfur activation about the hypocenter, owing to the tilt of the bomb at time of burst. Figure 9 shows the calculated and measured sulfur activation levels with the calculated values doing a good job of replicating the nonuniform spatial distribution of the measurements. Overall agreement is approximately 10% between calculated and measured values.

The neutrons which contribute to sulfur activation are only a small fraction of the total neutron fluence. Activation of europium and cobalt by thermal and epithermal neutrons is proportional to the total neutron fluence. Activation at all available sites at Nagasaki, as calculated

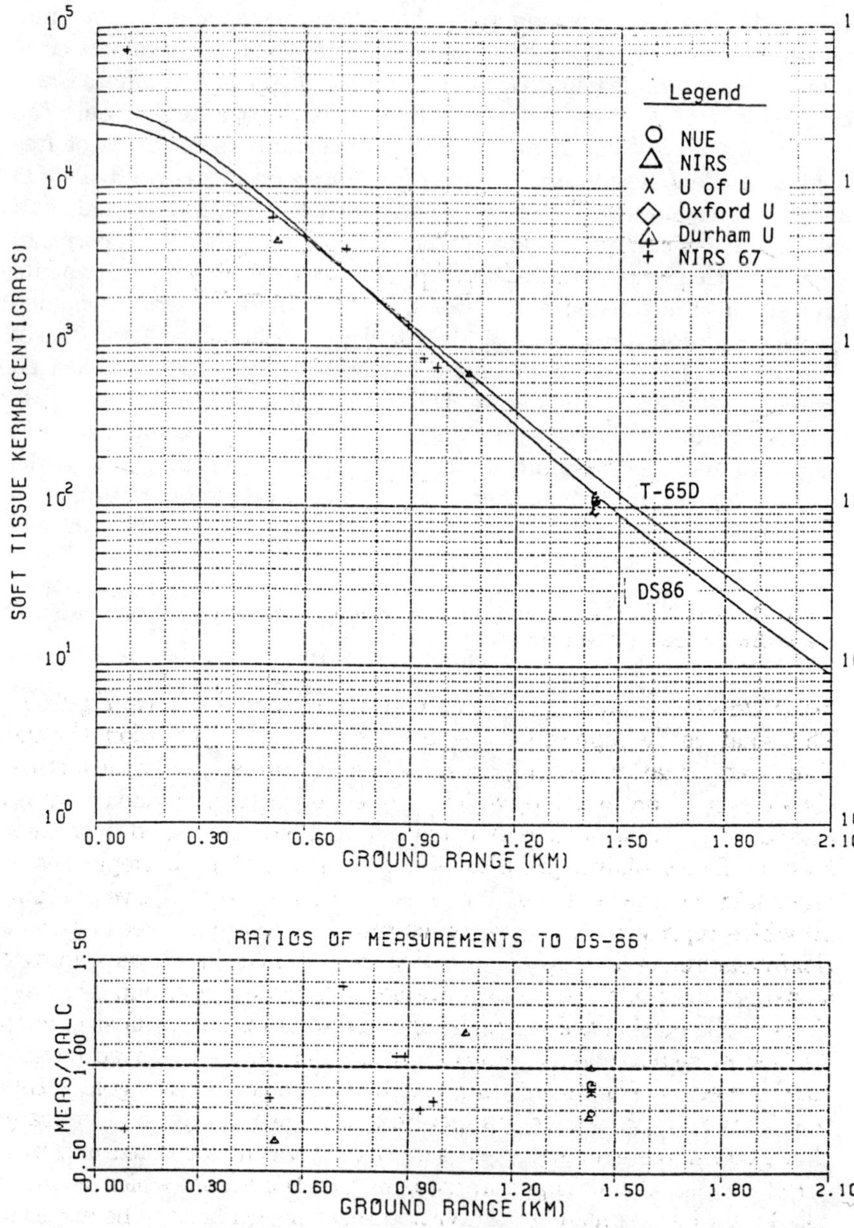

Fig. 6. DS86, T-65D and TLD-Measured Free Field Kerma Versus Ground Range at Nagasaki (New Map)

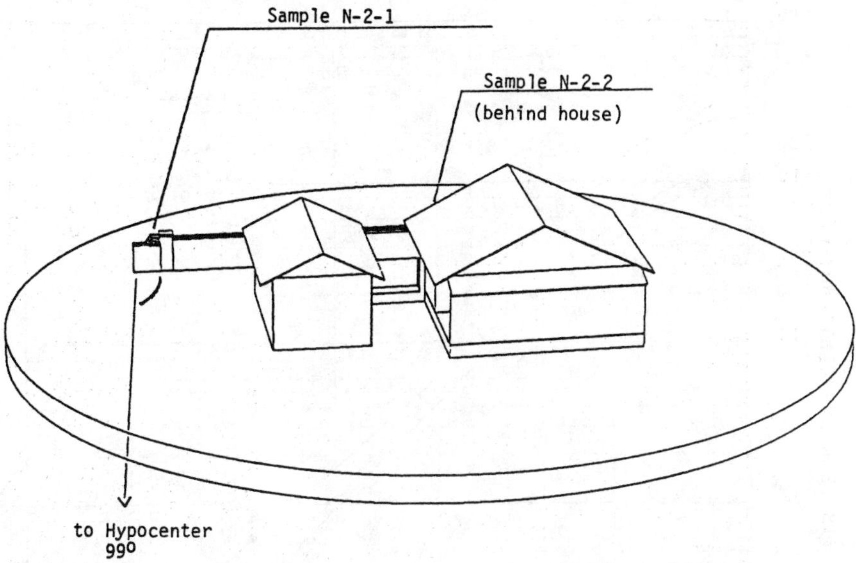

Fig. 7. Ieno Wall, House and Shed: TLD Sample Locations Approximately 1,435 Meters from Nagasaki Hypocenter

using DS86 methodology, is compared to measured values in Figure 10. The solid curve represents the best (subjective) regression through those data. That mean comparison is heavily influenced by the farthest cobalt activation location, which is itself uncertain, as shown by the attendant error bars. Use of transport methods improved over those used in DS86 shows promise of reducing existing discrepancies at Nagasaki. That is shown by the "revised methodology" curve in Figure 10 which represents a regression (subjective) through a set of calculation/measurement comparison ratios (not shown) based on improved transport methods. Those improvements include fine group cross sections in the epithermal energy range, reduced spatial mesh size, introduction of source energy group structure and revised delayed neutron source spectra. The use of those methods to calculate the gold activation from a similar source (Ranger Fox) as shown in Figure 11 indicates that good agreement has been achieved between calculated and measured values. The changes in transport methodology which result in this level of agreement for activation measurements may be expected to increase the free field neutron kerma at Hiroshima by approximately 6% and at Nagasaki by approximately 15% if implemented in DS86.

The results of calculation/measurement comparisons for europium and cobalt activation at Hiroshima, as shown in Figure 12, represent a

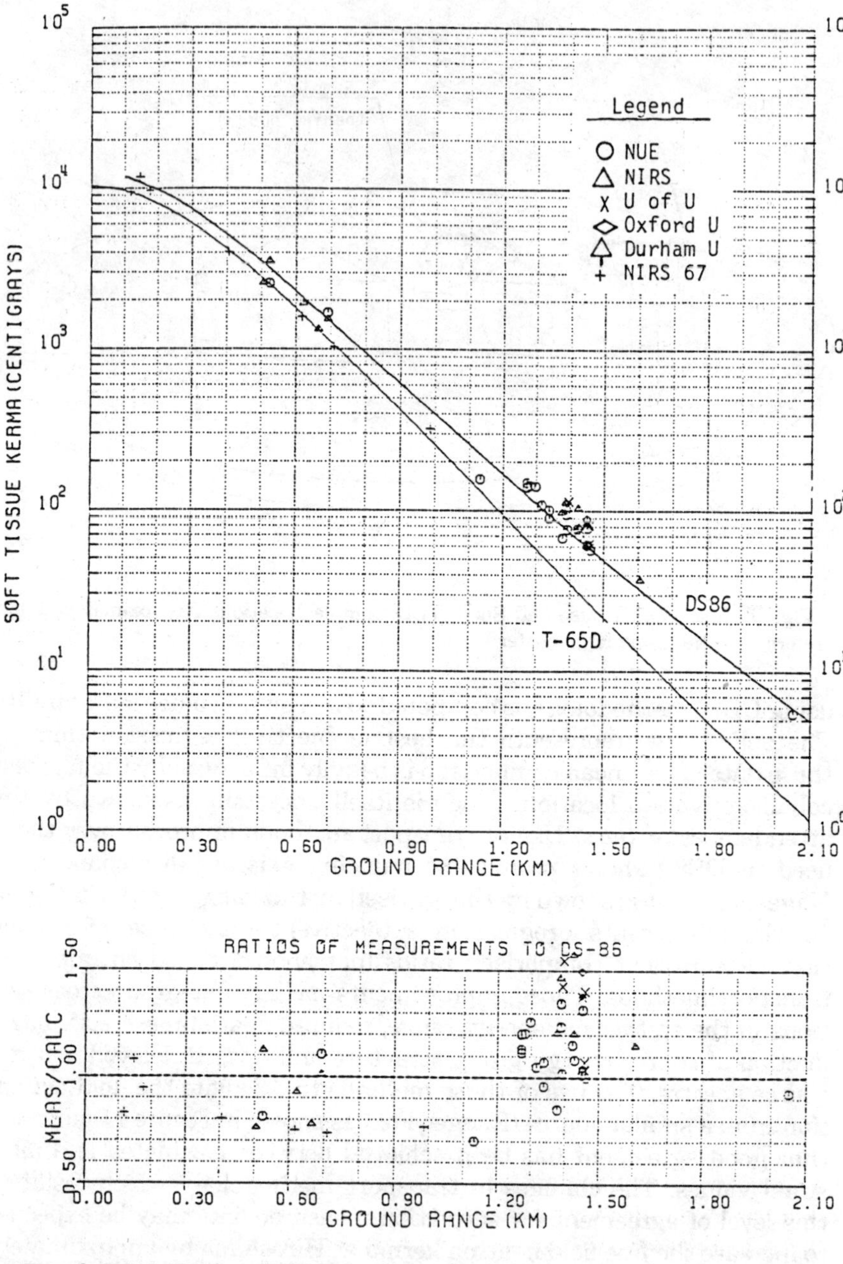

Fig. 8. DS86, T-65D and TLD-Measured Free Field Kerma Versus Ground Range at Hiroshima (New Map)

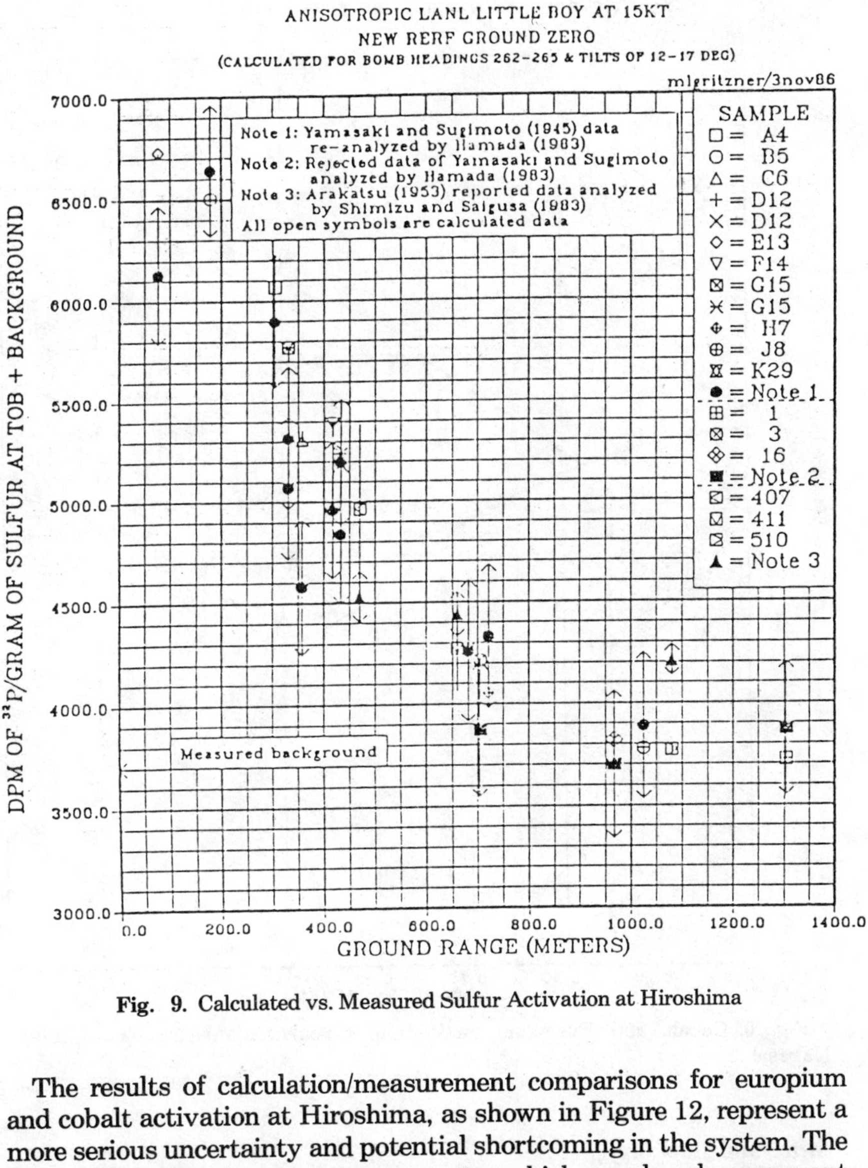

Fig. 9. Calculated vs. Measured Sulfur Activation at Hiroshima

The results of calculation/measurement comparisons for europium and cobalt activation at Hiroshima, as shown in Figure 12, represent a more serious uncertainty and potential shortcoming in the system. The transport methodology improvements which produced agreement between calculations and measurements for Nagasaki-like events have little impact on the observed discrepancy, which is of the order of a factor of five at 1,200 meters horizontal distance. This discrepancy and its implications for neutron fluence and dose uncertainty at Hiroshima are still under investigation.

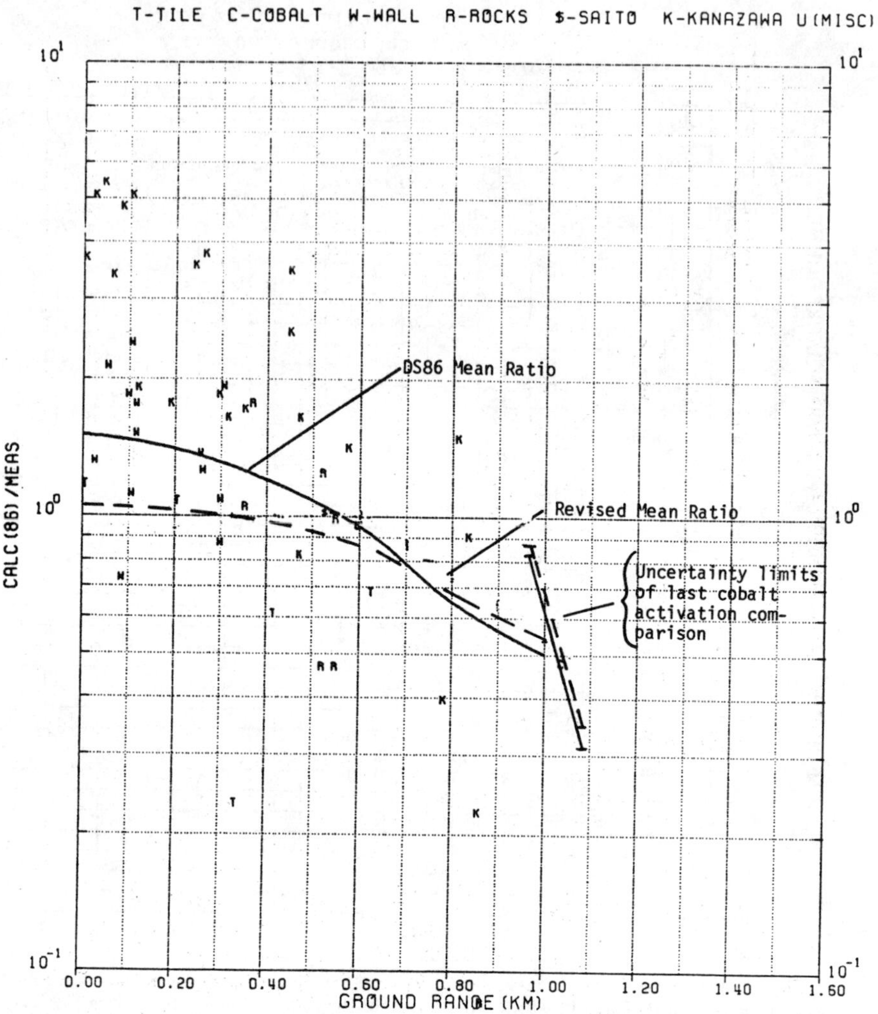

Fig. 10. Cobalt and Europium Activation; Calculation/Measurement Ratios at Nagasaki

Structural Shielding

The T–65D system depends primarily on two approaches to estimate survivor shielding. The first is the correlation of a survivor location with that of dosimeter location in a Japanese analogue house exposed at the BREN experiment, according to nine physical parameters developed from that experiment. The second is the assessment of the distri-

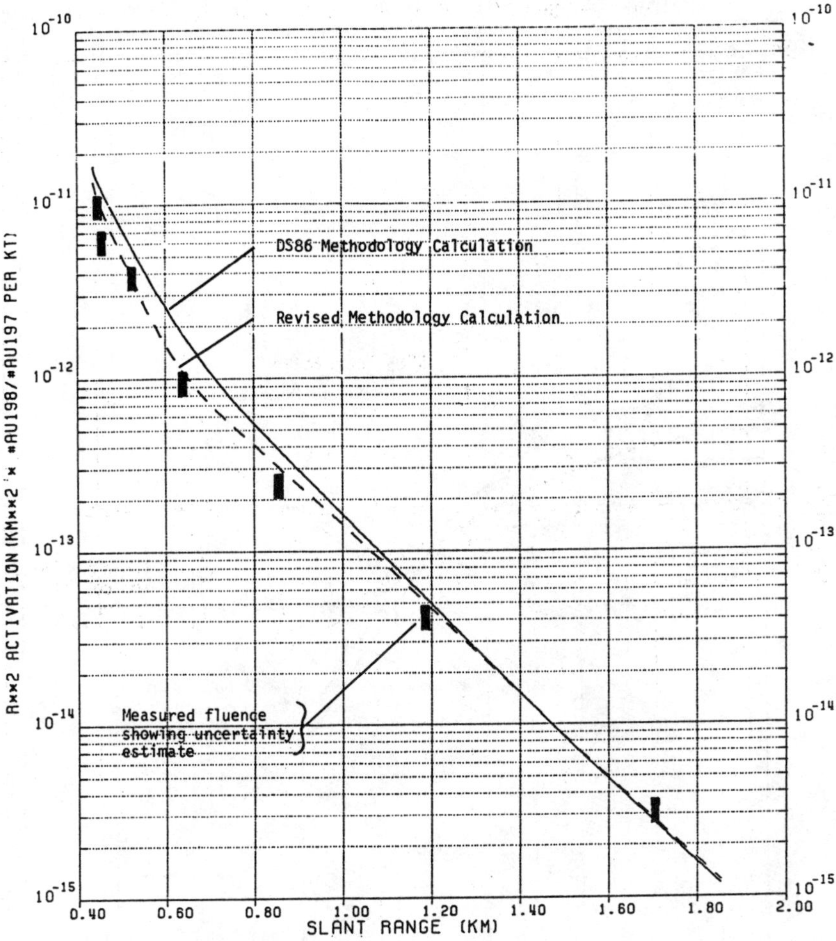

Fig. 11. Calculated and Measured Cadmium-Difference Gold Activation at Ranger Shot Fox as a Function of Slant Range

bution of unobstructed solid angle with respect to the source-target axis, the so-called Globe technique. The relative importance of each increment of solid angle is based on angular distributions of weapon radiation as measured at Nevada tests. Most survivors whose shielding is not assessed according to the above schemes but who are included in the follow-up study have been given an average transmission factor (shielded dose/free field dose).

Shielding data in DS86 are based on Monte Carlo calculations of radiation transport into locations in models representative of Japanese

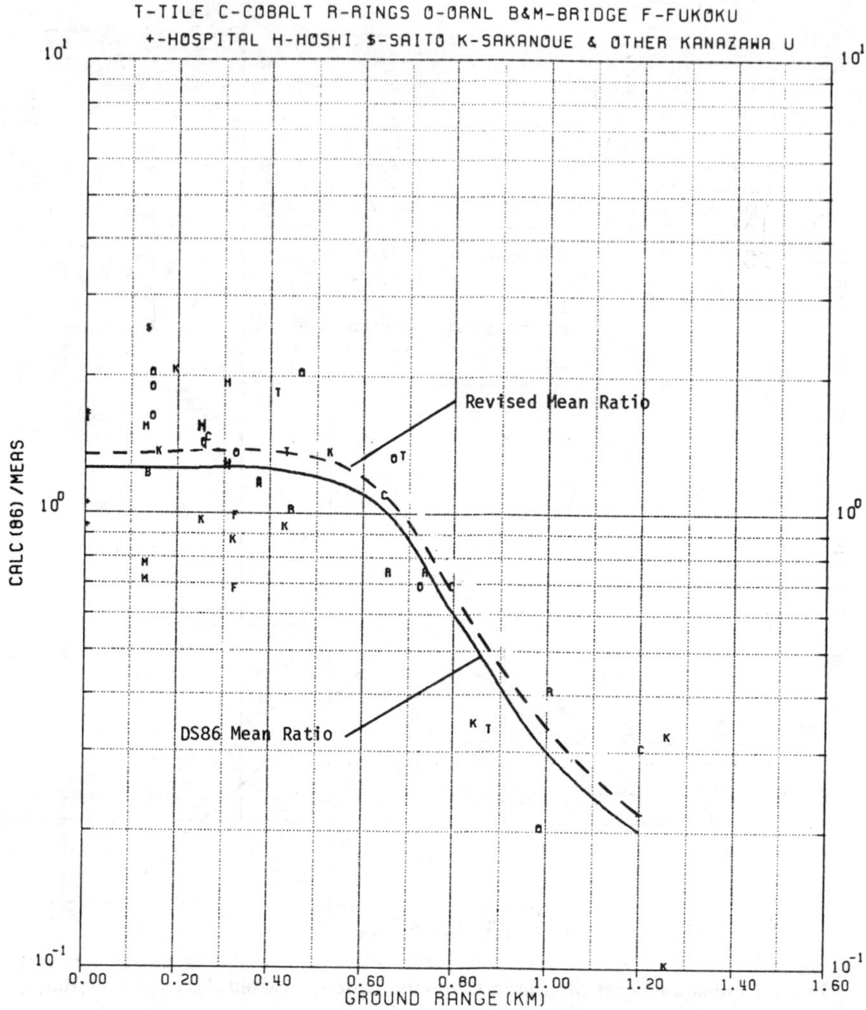

Fig. 12. Cobalt and Europium Activation; Calculation/Measurement Ratios at Hiroshima

neighborhoods. The DS86 transport methodology has been used to calculate transmission factors for Japanese analogue structures modeled after those exposed at the BREN experiment. Good agreement has been obtained between the BREN measurements and calculations of transmission factors for neutrons and for incident gamma rays in the absence of neutrons, as shown in Figure 13. In the presence of neutrons the measured transmission of gamma rays rises substantially above

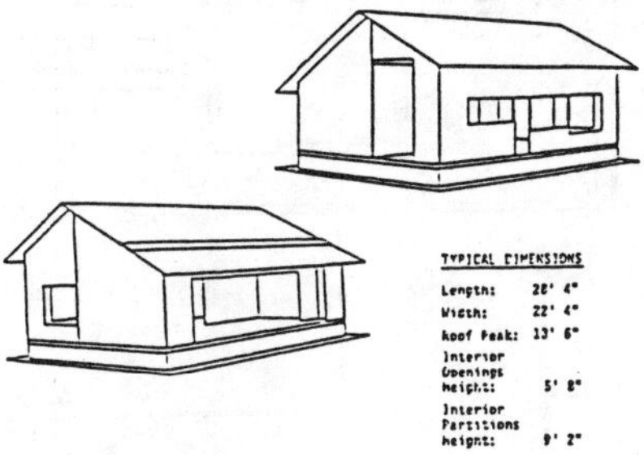

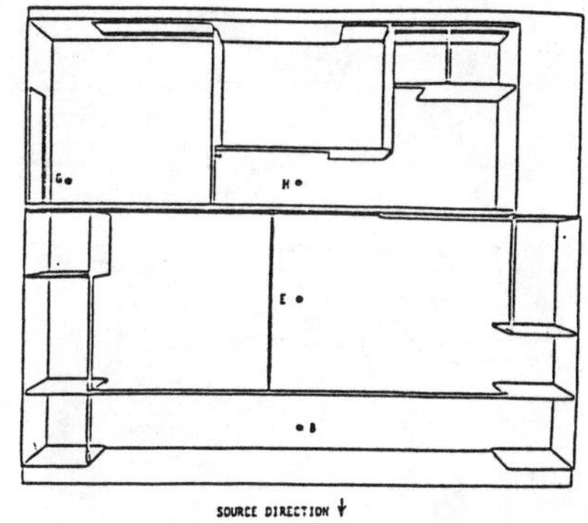

Geometry model for house Type A.

Interior of House A showing detector locations.

Location	Detector Height (ft)	^{60}Co Gamma Measured	^{60}Co Gamma Calculated	HPRR Neutron Measured	HPRR Neutron Calculated	HPRR Gamma Measured	HPRR Gamma Calculated
B	3	0.62	0.63	----	----	----	----
E	3	0.56	0.55	0.48	0.48	----	----
E	5	0.40	0.41	0.46	0.44	1.44	1.02
G	5	----	----	0.48	0.44	----	----
H	3	0.40	0.44	0.41	0.43	1.32	0.97
H	5	----	----	0.43	0.38	----	----

Fig. 13. BREN Shielding Experiment; House Type A Detector Locations; Measurements and Calculations of Radiation Transmission for Cobalt and Reactor (HPRR) Sources

the calculated values, however, those results are at odds with independent measurements made in the same houses at the subsequent Hardtack Test series, which give typical gamma-ray transmission values which are similar to those calculated for BREN, as shown in Figure 14. Unfortunately, the Hardtack data have not been matched by calculations for lack of weapon radiation source estimates. Finally, TLD measurements made at the Ieno wall location with and without an intervening house provide a transmission factor of 0.48 versus an equivalent calculated value of 0.52. Thus, the available measured data generally support the radiation transmission as calculated using DS86.

The quality of DS86 survivor shielding assessment is mainly affected by its limitation to shielding estimates based on the nine parameters of globe data as contained in the computerized survivor

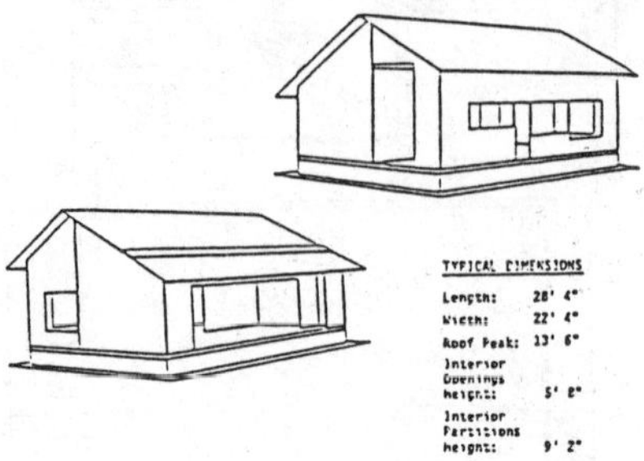

Geometry model for house Type A.

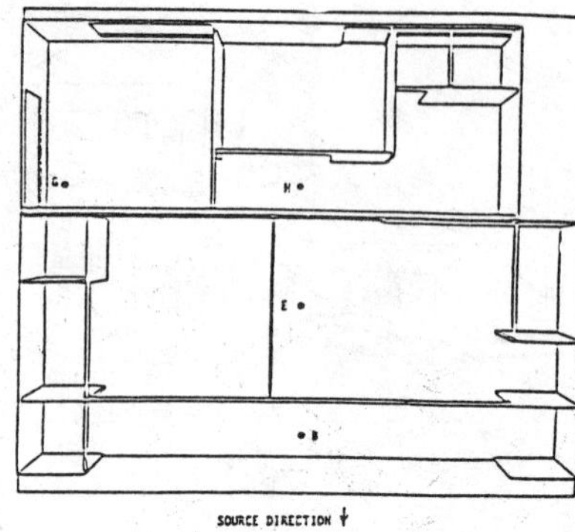

Interior of House A showing detector locations.

	Neutron Transmission			Gamma Transmission		
Location	BREN Measured	BREN Calculated	HARDTACK Measured	BREN Measured	BREN Calculated	HARDTACK Measured
5' E	0.46	0.44	0.41	1.44	1.02	1.08
3' H	0.41	0.43	0.35	1.32	0.98	0.98

Fig. 14. BREN and Hardtack Shielding Experiment; House Type A Detector Locations; Measurements and Calculations of Reactor Radiation Transmission; Measurements of Weapon Radiation Transmission

master file. Shortcomings in those data cause imprecise shielding assignments under DS86, such that the dose uncertainty due to shielding variability is the largest of all the uncertainty components in the system, being of the order of 10 to 20%.

A program of system testing against calculational benchmarks based on individual survivor shielding files is now underway and is intended to identify the major deficiencies associated with shielding assessments as currently implemented. That test program uses detailed information from the shielding files of several survivors to recreate whole neighborhoods as precisely as possible. An example of such a neighborhood is shown in Figure 15. Radiation doses calculated using those neighborhood models will be compared with those obtained for the same survivors using DS86. Those comparisons are intended to identify potential systematic bias in shielding assessment. They will

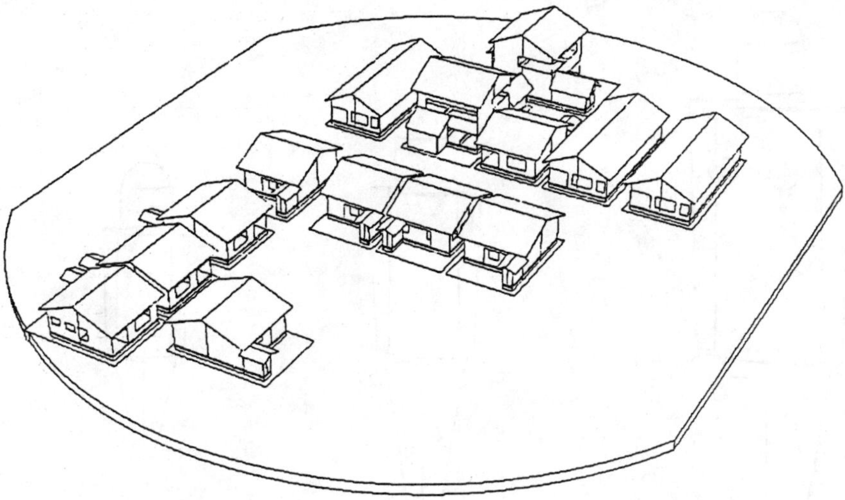

Fig. 15. Neighborhood of Residential Structures Drawn from Survivor Shielding Files and Used in Benchmark Testing of DS86

also help identify those types of data which might materially improve the precision of shielding assessment if retrieved from the individual shielding files, which are not computerized. Those parameters and their potential benefit will be assessed with great care because of the high cost of retrieving such information from the survivor shielding files.

Organ Dosimetry

Monte Carlo calculations of organ doses from incident neutrons and gamma rays have been performed using anthropomorphic phantoms representing Japanese male and female adults, children and infants, as shown in Figure 16. DS86 contains radiation transmission data for 15 organs in each phantom, as portrayed in three postures, standing, kneeling and prone, and in 45 degree orientation increments.

Few measurements have been made of depth doses in such phantoms exposed to radiation fields analogous to those present at Hiroshima and Nagasaki. Measurements performed in a small (3.5 Kg) tissue phantom to determine dose from incident neutrons and gamma rays from a TRIGA reactor are generally within 10% of matching calculations, as shown in Table 2. Results will soon be available from dose measurements made in an anthropomorphic adult phantom exposed to neutrons and gamma rays 400 meters from a bare fast reactor. Those

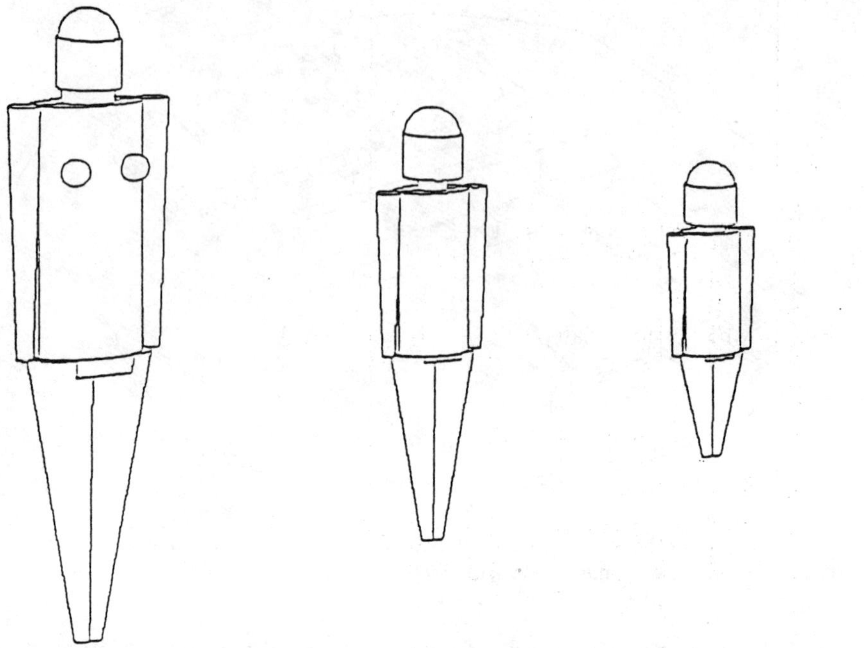

Fig. 16. External Configuration of Three Phantoms Representing Adult, Child and Infant Survivors

measurements will provide the best test yet of radiation transport calculations in phantoms.

Dose Uncertainty

DS86 contains a simple system for the assessment of uncertainty of organ doses in individual survivors. It assigns percent standard deviations to quantities calculated in the system. Those percent standard deviations have been estimated, sometimes subjectively, based on the work of the individual contributors to DS86 and are shown in Table 3. System computed variances are combined to obtain the variance on the dose from each radiation component and on the total. It has been an objective of the program to obtain percent standard deviations of 25 or

TABLE 2—*Experimental and Calculated KERMA — Exposure Room 1, AFRRI TRIGA Reactor*

Configuration 90 cm from core midline, 2″ pb intervening

Location	Component	EXP. Free Field KERMA (Rads per kw-min)	PHANTOM KERMA (Rads per KW-min) EXP. (FSD)	Calc. (FSD)	Δ(1)	TRANSMISSION EXP	Calc.	N/γ Ratio EXP	Calc.
Mid-Head	n	10.1(.03)	5.69(.04)	5.49(.05)	–.04	0.56	0.54		
	n–γ			0.56(.07)			0.06(2)		
	γ			2.46(.04)			0.95		
	TOTγ	2.6(.03)	3.35(.03)	3.02(.04)	–.10	1.29(3)	1.16		
	TOT	12.7(.03)	9.04(.03)	8.51(.03)	–.06	0.71(4)	0.67	1.70	1.82
Mid-Thorax	n	11.8(.03)	7.1(.04)	6.33(.05)	–.11	0.60	0.54		
	n–γ			.98(.06)			0.08		
	γ			2.83(.04)			0.91		
	TOTγ	3.1(.03)	4.1(.03)	3.81(.04)	–.07	1.32	1.23		
	TOT	14.9(.03)	11.2(.03)	10.14(.03)	–.09	0.75	0.68	1.73	1.66

(1) $\frac{\text{CALC.–EXP}}{\text{EXP}}$

(2) $\frac{\text{n–}\gamma\text{ Phantom KERMA}}{\text{n Free Field KERMA}}$

(3) $\frac{\text{Total }\gamma\text{ Phantom KERMA}}{\text{Total }\gamma\text{ Free field KERMA}}$

(4) $\frac{\text{Total Phantom KERMA}}{\text{Total free Field KERMA}}$

TABLE 3—*Uncertainty Contribution Estimates by Model and Radiation Component for DS86 (Percent Standard Deviation)*

Model	Radiation Component: Prompt Neutron	Delayed Neutron	Early Gamma	Late Gamma	Neutron-Induced Gamma
Yield (Hiroshima)	10	12	10	12[1]	——SIMILAR TO NEUTRON PSD's——
Yield (Nagasaki)	5	6	5	6	
Neutron Output	5	–	5	–	
Air Density	5	7	3	3	
Air Moisture	5	5	1	1	
Hydro Model	–	10	–	10	
Fission Source	–	20	–	10	
Air Transport Calculations	5	20	5	10	
Hypocenter Location	10	10	6	4	
House Location	5	5	3	2	
House Materials	5	5	5	5	5
Survivor Location	10–20(30)[2]	10–20(30)	10–20(30)	10–20(30)	15
Shielding Calculation Method	5	5	5	5	10
Man Materials	5	5	5	5	10
Man Orientation	10–15(25)	10–15(25)	8–12(15)	8–12(15)	5–15
Body Shielding Calculation Method	5	5	5	5	5
Hiroshima	25–32(44)	39–44(54)	22–29(38)	27–33(41)	29–32
Nagasaki	23–31(43)	38–43(53)	20–28(37)	25–32(40)	27–31

1. Most important uncertainty contributions or those most open to question.

2. Values in parenthesis () apply to unknown survivor shielding

better on the doses for the best characterized survivors. As can be seen from Table 3 we have achieved that goal in part.

Those percent standard deviations most open to question and those which have the most impact on survivor dose are highlighted with boxes in Table 3. It is possible that the Hiroshima yield uncertainty can be reduced to that for Nagasaki through the completion of further analysis of the canister and I-beam data. The assigned neutron transport uncertainties may be overly optimistic, based on the evidence of activation measurements, and may have to be increased to 200 to 1,000 percent, depending on distance, all in the upward direction, pending resolution of the activation problem. Survivor location (shielding assignment) uncertainties may be reduced based on the findings of the benchmark testing program, with reductions of as much as a factor of two possible for the best characterized survivors. Uncertainties in the organ dose due to survivor orientation variation are not likely to be improved beyond those given. Altogether, forseeable improvements should reduce the aggregate uncertainty of the total organ dose and that from each radiation component below 25 percent standard deviation for the best characterized survivors, with the possible exception of neutron dose.

Summary

The DS86 dosimetry system has been extensively tested against benchmark measurements and calculations to determine its accuracy. The agreement between DS86 and benchmark data is generally very good. Therefore, persons interested in using DS86-based dose data for epidemiological studies should do so secure in the knowledge that DS86 represents a substantial improvement over T–65D, the system that it supercedes.

There are some outstanding issues, the resolution of which may cause modifications in the current system at some time in the future. The most important of those issues pertains to uncertainties in the neutron dose at Hiroshima and the precision with which shielding categories are assigned to individual survivors. Those and other issues are being pursued on a continuing basis by a technical group headed by Dr. Robert Christy, under funding provided by the U.S. Department of Energy.

Discussion

K. CLIFTON: When I was in Hiroshima in the early 80's before the T65D blew up so to speak, there was some biological evidence of an asymmetry in the isodose curves at the ground level. There was some indication that the isodose curves might be elliptical rather than circular and I was wondering whether this is something of real concern. This was thought to be primarily true in Hiroshima because of the orientation of the bomb at the time of its detonation.

D. KAUL: the asymmetry of the "footprint" of the fast neutrons, which are the only component thought to be so affected, is not part of DS86. DS86 assumes a symmetrical distribution. But the inclusion of that would not be impossible, merely very difficult. I've been asked to mention the fact that it was taken into account in the analysis of the sulphur, but it is not part of the DS86 *per se.*

R. CHRISTY: In that same connection, I might add that I was puzzled at one point to find in the biological data at Nagasaki, particularly some of the cancer incidence, some evidence of asymmetry which I didn't think could have readily arisen from the bomb. It appeared that the population characteristics in Nagasaki were somewhat different sectors and I was informed that this could well have accounted for the asymmetry in the biological effects. I think when you observe peculiarities in biology, you have to look at the peculiarities of the subjects as well as of the radiation.

F. SEITZ: Were these differences ethnic or dietary?

R. CHRISTY: Ethnic, I would suppose. Seymour has the answer.

S. JABLON: Probably mostly socio-economic.

W. SINCLAIR: Can I ask a different kind of question? I think you referred to both the europium and cobalt uncertainties as being relevant to our future problems, if you like. And I understand the cobalt uncertainty very well in its magnitude. But I thought Dr. Kosako showed very good agreement between the measurements and the calculations in the case, but in that case you'll notice that out at 1200 m the europium measurements were well above that particulare regression, or that particular line.

D. KAUL: The curves that Dr. Kosako had drawn through his data were, I believe, regressions through the data. They had nothing to do with DS86. Well in one case, but in that case you'll notice that out at

1200 m the europium measurements were well above that particular regression, or that particular line.

W. SINCLAIR: Well, perhaps could you give us in a word what the magnitude or the status of the europium problem is?

D. KAUL: At Hiroshima, inside about 700 m the data indicate that the calculations are high. Beyond about 700 m the calculations are uniformly low, and that includes a number of different sources of data. That includes the europium measurements from several laboratories in Japan. That includes the cobalt measurements not only made by the Japanese but also by Oak Ridge. Now the Oak Ridge data has a difficulty in that there is no absolute normalization to it, but if you normalize it to one of the highest precision ring locations, external locations which the Oak Ridge data are for, external iron activation, they line right up with the other measured data.

W. SINCLAIR: So we actually have a very similar problem?

D. KAUL: We have, I think, high resolution of the problem that exists beyond 700 m.

J. PACE: Dean, I'd like to ask you a couple of questions. (1) Could you tell me something about the revised methodology, and (2) how much would 1000 percent improvement in that low energy cross section change the dose to an individual?

D. KAUL: The revised methodology: We found at Nagasaki, that the problem that exists with the current comparisons inside 600 m or so, could be laid at the feet of the multi-group cross sections that were used to transport the Nagasaki device source. Those cross sections are designed primarily to transport high-energy neutrons, which upon scattering rapidly achieve a 1/E spectrum. And those cross sections tend to be dominated in the broad group weighting by a 1/E shape. We found, that when we used very fine group cross sections the vitamin E, 174 group set, that the low-energy neutrons, those below 3 keV, which are the vast majority of neutrons from the Nagasaki device, indeed don't go anywhere. They go less somewhere than even DS86 predicts. So, basically replacing the broad group cross sections with fine group cross sections below a few keV, changed the comparison at Nagasaki, but as you saw, not at Hiroshima, because it's not similarly dominated by those neutrons. the calculation, from 25 m, which it is for DS86, to 5 m or so. Because these transport calculations are so dominated by what are considered, by current standards, to be lower energy neutrons, the use of a longer mesh size tends to cause an extrapolation in the calculation which underestimates the dose. And cutting the mesh size down to 5 m or so, changes the dose by about 15 percent out beyond a km. Now as to your other comment: That same change in mesh size is going to similarly affect kerma. It's not going to give you all of the 15 percent but it's going to give you some of it. As far as the impact of the uncertainties, as I said, the source and transport range that creates the cobalt activation is between about 3/4 and 1 1/2 MeV. For tissue kerma it's between 1 1/2 and 2 1/2 MeV. For sulphur, it's a little over 2 to well over 6 MeV. Therefore, the tissue kerma seems to be more closely tied to the colbalt and europium activation than it is to sulphur; at least in that region where neutrons are likely to be important if they are increased by any significant percentage.

Perspectives on Special Aspects of the Dosimetry — Panel Discussion

Chairman, **William H. Ellett**
National Research Council

John H. Harley
Hoboken, New Jersey

George D. Kerr
Oak Ridge National Laboratory

William E. Loewe
Lawrence Livermore Laboratory

Paul P. Whalen
Los Alamos National Laboratory

William H. Ellett

Each of the panel members with me today has made a unique contribution to the dose reassessment. Dr. William Loewe's and Edgar Mendelsohn's initial publication in Health Physics of A-bomb doses based on Preeg's 1976 source spectrum for the Hiroshima weapon was a seminal event in the long series of investigations that lead to the new A-bomb dosimetry. Dr. Loewe's interest in these problems has been a continuing one, and he has been a leader in the comparison of calculated neutron kermas to kerma estimates based on experimental data obtained in Japan.

Dr. George Kerr is a bulwark in the dose reassessment activities. He established the meteorologic and geographic data bases for the bombings, and, along with Joe Pace, independently verified the results of Loewe and Mendelsohn's studies. Their subsequent calculations of neutron and gamma-ray transport in the two cities are the bases of the new dose estimates. Dr. Kerr is a large part of the U.S.A.'s institutional memory on the atomic bombing and the subsequent efforts at Oak Ridge and elsewhere to understand what happened in the two cities.

The source spectrum for the Hiroshima and Nagasaki bombs that are the starting point of the new dose estimates were calculated by Paul Whalen and his co-workers at Los Alamos. Paul also took a very active part in the organization of an experimental program to measure neutron fluences from the replica of the Hiroshima weapon that was assembled at Los Alamos. He also calculated the output spectrum for the replica and made the comparisons of experimental and calculational results that verified the source spectrum in the dose reassessment.

A continuing question in the discussion of the A-bomb survivors is what role fallout and activation products may have had in the exposures that occurred in the two cities. This problem was identified by the Seitz panel and Dr. John Harley was asked to explore this question. After reading many old Japanese and U.S. reports Dr. Harley and his Japanese collaborators were able to define the roles of these sources of exposure in terms of both space and time. It is such a thorough job that it is unlikely that this problem will be revisited.

We have already heard several overviews of the dose reassessment. For this panel, whose members worked so diligently on the nuts and bolts of the study, a more personal viewpoint is perhaps appropriate. I have asked each of them to tell you about that aspect of their own investigations which they found most interesting and to explain how it fit into our understanding of A-bomb dosimetry.

Dose From Residual Radioactivity at Hiroshima and Nagasaki

John H. Harley
Hoboken, New Jersey

Abstract

The majority of the radiation dose delivered at Hiroshima and Nagasaki came within minutes of the detonations. A smaller dose was delivered over an extended period from residual radioactivity. There were two sources, the radioactivity induced in soil and other materials near the hypocenters and the fallout of fission product radionuclides from the weapon clouds.

Induced radioactivity was comparable at the two cities but the fallout at Nagasaki was much more intense than at Hiroshima. The maximized integrated absorbed doses in tissue were about 50 rad and 20 rad for induced radioactivity and 1 rad and 20 rad for fallout, respectively, in Hiroshima and Nagasaki. Actual dose contributions should be smaller and are of little significance in the dose reassessment except that those few individuals involved are poor candidates for inclusion in the control population.

Doses from Residual Radioactivity in Hiroshima and Nagasaki

In the previous assessments of radiation exposure from the nuclear explosions at Hiroshima and Nagasaki, no attempt was made to evaluate the contribution from residual radioactivity. In the present report, estimates have been made for both the exposure from radioactivity induced in soil at the hypocenter and that from fallout of fission products following the explosions.

There were reports at the time of "black rain" occurring at sites about 3 km from the hypocenters. In Hiroshima, this fallout appeared in the Koi-Takasu area and, in Nagasaki, in the Nishiyama area. Radiation measurements made about two months later showed elevated gamma exposures in these areas. These have not been considered significant compared to the overwhelming exposure of the population from direct radiation.

The present work attempted to collect all of the disparate data and to bring the information together in a coherent form. The exposure rate at one meter height at an arbitrary time was selected as a common point. The original Japanese measurements appear to be generally sound and they were supplemented by measurements made by teams from the Naval Medical Research Institute (NMRI) and the Manhattan Engineering District (MED). The instruments included Lauritsen electroscopes, Neher cosmic-ray chambers and portable Geiger counters. Fortunately, all of these were calibrated against radium sources and the exposure rates are comparable. While we wish in retrospect that more data had been collected, the measurements were a remarkable achievement under the existing conditions.

There are four regions of interest, the hypocenter and the fallout area for each of the cities. The radionuclides involved and their radiation properties are very different for the two sources and will be treated separately. The hypocenter conditions were similar at Hiroshima and Nagasaki but the fallout was more intense near the latter city.

Fission Product Fallout

The "black rain" observed in Japan has never appeared in the tests of similar weapons carried out at the arid Nevada Test Site. Several sources of water may have contributed, but the most likely are water drawn up into the cloud from the various bodies of water under the explosion and the high humidity existing in the atmosphere of Japan in August. The rainout occurred from one-half to one hour after the explosion and continued for an hour or two.

The fission products from a nuclear explosion are a complex mixture with a wide range of half-lives and radiation energies. In general, the gamma-ray output may be considered to decay according to a power law with an exponent of -1.2. Thus, the measurements taken at any time may be extrapolated to a common time or the cumulative exposure from the deposition time to any other time may be estimated. In this study, the cumulative exposure from one hour to infinity was calculated. Table 1 shows the way that exposure accumulates with time when this process is followed.

The cumulative exposures for Hiroshima and Nagasaki that would be estimated from the data of different investigators are shown in Table 2. The agreement is reasonable, and it is apparent that fallout in Hiroshima did not make a significant dose contribution.

TABLE 1—*Effect of Time of Entry into a Fallout Area Having an Exposure Rate of 1 R/h at One Hour after the Explosion. The Fission Product Decay is Calculated as* $t^{-1.2}$

Time of Entry	Exposure Rate (R/h)	Cumulative Exposure (R) from Entry to Infinity
1/2 h	2.3	5.7
1 h	1.0	5
2 h	0.44	4.4
4 h	0.19	3.8
8 h	0.082	3.3
1 day	0.022	2.6
1 week	0.0021	1.8
1 month	0.0004	1.3

TABLE 2—*Estimates of Exposure Rate at 1 Hour and Cumulative Exposure at 1 Meter Height from Measurement of Radioactive Fallout.*

Investigator		Exposure Rate at 1 h (R/h)	Cumulative Exposure (R)
NAGASAKI (Nishiyama)			
Shinohara, et al. (1953a)	Range	0.5–14	2.5–70
MED[a]	–	5.8	29
MED[b]	Max.	4.7–8.5	24–43
NMRI[c]	Max.	8.4	42
	Village	7.6	38
HIROSHIMA (Koi-Takasu)			
Miyasaki and Masuda (1953)	Max.	0.6	3
	Average	0.5	2.5
Fujiwara and Takeyama (1953)	Max.	0.19	1
MED[d]	Max.	0.23	1.2
NMRI[c]	Range	0.11–0.33	0.6–1.6

a) Tybout, Quoted by Wilson (1956)
b) Reported by McRaney and McGahan (1980)
c) Reported by Pace and Smith (1958)
d) Tybout, quoted by Arakawa (1962)

An additional check on the Nagasaki exposure comes from measurements made on soil samples collected in 1956, before the heavy global fallout from thermonuclear testing. Miller (1982) reported ^{137}Cs measurements on soil from Nishiyama and from an area outside the fallout zone as 62 and 7 mCi/km^2, respectively. The difference, extrapolated back to 1945 and adjusted for the presence of the other fission products, gave an integrated exposure of about 40 R. It should be noted that global fallout from all weapons tests amounted to about 100 mCi/km^2 of ^{137}Cs.

The data in Table 2 relate to relatively small areas of maximum fallout. Radiation levels fall off with distance, although the contours at Nagasaki indicate that an exposure of about one-fifth the maximum covered an area of about 1,000 hectares. The actual exposure of individuals can only be developed if a record of their day-by-day movements is available for, say, a month after the explosion. Okajima (1985) made an estimate for a group of Nishiyama residents and concluded that their actual exposure was about two-thirds of the 40 R estimated above.

The exposures from inhaled and ingested fission products are much more difficult to estimate. Qualitatively, the inhalation route is unlikely to be significant. For ingestion, Okajima (1982, 1987 NAS Report) estimated that the contribution from ^{137}Cs would be about 10 mrem over the years from 1945 to 1985. This is not large, but the contributions from the short-lived radionuclides, such as ^{131}I, were probably much higher.

In summary, the external gamma radiation exposures from fallout appear to be in the range of 20 to 40 R for Nagasaki and 1 to 3 R for Hiroshima when integrated from 1 hour to infinity.

Induced Radioactivity

The residual radioactivity near the hypocenter results from radionuclides produced by neutron irradiation of the soil and building materials. The identity and amount of these nuclides depends on the composition of the soil as well as the neutron fluence. With the soil composition involved, the radionuclides shown in Table 3 are expected to be the most significant.

It is apparent that most of the radiation exposure comes from the short-lived radionuclides that would have disappeared before the measurements made at about two months after the explosions. Thus, it was necessary to estimate the radionuclide composition at earlier

TABLE 3—*Induced Radionuclides of Possible Dosimetric Interest.*

Nuclide	Half-Life	
Al-28	2 m	
Mn-56	2.6 h	
Si-31	2.6 h	Not significant
K-42	12.4 h	Not significant
Na-24	15 h	
Fe-59	44.5 d	Not significant
Sc-46	83.8 d	
Cs-134	2.1 a	
Co-60	5.3 a	

times from neutron activation analyses of soils made later. In the calculations reported here, the data of Hashizume et al. (1970) for Hiroshima were used for both cities. The relevant values from Hashizume are shown in Table 4. A composite decay curve was drawn for the mixture of radionuclides and this was used to estimate exposure rates and cumulative exposures.

Since the areas at the hypocenters were completely destroyed and intense fires were burning, it was considered that no one entered the areas earlier than one day after the explosions. This limited the contributing radionuclides to ^{24}Na, ^{46}Sc, ^{134}Cs and ^{60}Co.

The cumulative exposure estimates from the various measurements made in the two cities are shown in Table 5. The most likely values are about 80 R for Hiroshima and 30-40 R for Nagasaki. These exposures fall off with time and distance from the hypocenter. The cumulative exposure would be about one-third as large after a day and only a few percent after a week. As an example, an individual entering the Hiroshima hypocenter area after one day and working 10 or 12 hours

TABLE 4—*Exposure Rates Estimated by Hashizume et al. (1970) for Hiroshima Soil at the Hypocenter for t = 0 and the Cumulative Exposures for Complete Decay.*

	Initial Exposure Rate (R/h)	Cumulative Exposure (R)
Mn-56	4.8	18
Na-24	3.0	64
Sc-46	8.5×10^{-5}	0.25
Co-60	1.1×10^{-5}	0.73
Cs-134	4.8×10^{-9}	0.0001

Note that the actual exposure rate at t = 0 would not be the sum of these individual longer-lived radionuclides.

TABLE 5—*Estimates of Initial Exposure Rates and Cumulative Exposures from Total Induced Radioactivity Near the Hypocenters.*

Investigator	Exposure Rate at t = 0 (R/h)	Cumulative Exposure (R)
HIROSHIMA		
Hashizume et al. (1975)	8	80
Arakawa (1962)	2*	24*
Takeshita (1975)	5	98
Shohno (1965)	–	130
Gritzner (1986)	8	100
This report	9	70
NAGASAKI		
Hashizume et al. (1975)	–	30
Arakawa (1962)	0.8*	4*
Takeshita (1975)	-	39
Shohno (1965)	-	55
Gritzner (1986)	12	35

* Indicates values stated by Arakawa to be the best of the 4 models tested by him. They are based on neutron fluence estimates that are lower than those given in the NAS report by a factor of 5 for Hiroshima and of 2 for Nagasaki.

a day for a week would have been exposed to about 10 R. If he had been working at 0.5 km, the exposure would have been 1 R, and at 1km, about 20 mR.

Some measurements were made on the composition of bricks and tiles. Most noteworthy was the manganese content of tiles, which were higher than soil by a factor of five. The short-lived ^{56}Mn would not have been measured and its contribution to exposure cannot be estimated at this time.

The expected exposures near the hypocenters were calculated (Gritzner and Woolson, 1986; Gritzner, 1986) from the neutron fluences and soil compositions estimated for the NAS report. The results are in reasonable agreement with those described previously and are listed in Table 5.

Summary

This report has presented available data and calculations for assessing the exposure of survivors in Hiroshima and Nagasaki to external gamma radiation from residual radionuclides. The internal contribution from long-lived fission products appears small but there is no way to estimate potential exposures from ingested short-lived fission products.

For comparison with organ doses estimated in other parts of the NAS report, the exposure results have been converted to absorbed dose (assumed equal to kerma) in tissue. A factor of 0.87 is used to convert R to rad in air and the UNSCEAR (1982) factor of 0.7 is then used to convert to rad in tissue for the whole body. This factor accounts for the change in material (air to tissue) and for back-scatter and shielding by tissues of the body. No allowance has been made for shielding by buildings.

The cumulative fallout exposures given in this report become absorbed doses in tissue of 12-24 rad for Nagasaki and 0.6-2 rad for Hiroshima. The corresponding doses from induced radioactivity become 18-24 rad and about 50 rad, respectively.

These estimated doses were delivered to relatively few people. Population size estimates are lacking except for the report of Okajima (1975), listing the population of Nishiyama as about 600. The individual exposures from residual radiation may not be significant compared with the direct radiation at the time of the bombings. On the other hand, individuals with potential exposure from this source are dubious candidates for inclusion in a non-exposed cohort.

Acknowledgement

This report is a shortened and simplified version of the chapter on residual radiation prepared for the main NAS report. The chapter was written by Shunzo Okajima, Shoichiro Fujita and John Harley with the support of Eizo Tajima and Takashi Maruyama.

References

1. Arakawa, E.T. (1962) Residual radiation in Hiroshima and Nagasaki Technical Report 02-62. Hiroshima: ABCC.
2. Fujiwara, T., and H. Takeyama (1953) 22. Residual radioactivity around Hiroshima city. In: Collection of reports on investigations of the atomic bomb casualties. Tokyo: Science Council of Japan, pp. 75-83.
3. Gritzner, M.L. (1986) Unpublished Calculations for this Chapter. Science Applications International Corporation. La Jolla, CA.
4. Gritzner, M.L., and W.A. Woolson (1986) Calculation of A-Bomb Induced Soil Activation at Hiroshima and Nagasaki. Report SAIC-86/1503. Science Applications International Corporation. La Jolla, CA.

5. Hashizume, T., and T. Maruyama (1975) Dose estimation from residual and fallout radioactivity. 2. A simulated neutron activation experiment. J. Radiation Research 16 : 32-34.
6. Hashizume, T., T. Maruyama, Y. Kumamoto, Y. Kato, and S. Kawamura (1970) Estimation of gamma-ray dose from neutron-induced radioactivity in Hiroshima and Nagasaki. Technical Report 16-70. Hiroshima: ABCC.
7. McRaney, W., and J. McGahan (1980) Radiation dose reconstruction — U.S. occupation forces in Hiroshima and Nagasaki, Japan, 1945-46. Report DNA 5512F. Washington, DC: Defense Nuclear Agency.
8. Miller, K. (1982) memorandum to Wayne Lowder, September 14, 1982. Environmental Measurements Laboratory, New York.
9. Miyasaki, T., and T. Masuda (1953) A-bomb radiation in Hiroshima City and vicinity (2). In: Collection of reports on investigations of the atomic bomb casualties. Tokyo: Science Council of Japan, pp. 35-38.
10. Okajima, S. (1975) Fallout in the Nagasaki-Nishiyama district. J. Radiation Research 16-Supp : 35-41.
11. Okajima, S. (1985) Exposure of Nishiyama residents in Nagasaki to radioactive fallout-Investigation of behavior immediately after the A-bomb. Unpublished.
12. Okajima, S., J. Miyajima, I. Morimoto, S. Kawamoto, and T. Ishimaru (1982) Effects of the radioactive fallout of the Nagasaki atomic bomb. Hiroshima Igaku (in Japanese) 35: 334-36.
13. Pace, N., and R.E. Smith (1959) Measurement of the Residual Radiation Intensity at the Hiroshima and Nagasaki bomb sites. Technical Report 26-59. Hiroshima: ABCC.
14. Shinohara, K., A. Morita, K. Kora, and M. Yokota (1953) 14. Radioactivity of the ground in Nagasaki City and vicinity. In: Collection of reports on investigations of the atomic bomb casualties. Tokyo: Science Council of Japan, pp. 45-53.
15. Shohno, N. (1965) On residual radiation. J. Hiroshima Med Assn-Supplement 20(1) : 75-91.
16. Takeshita, K. (1975) Dose estimation from residual and fallout radioactivity. 1. Areal surveys. J. Radiation Research 16-supp : 24-31.
17. UNSCEAR (1982) Exposures resulting from nuclear explosions. In: Ionizing Radiation: Sources and Biological Effects. New York: United Nations, pp. 249-332.
18. Wilson, R.R. (1956) Nuclear radiation at Hiroshima and Nagasaki. Radiation Research 4 : 349-359.

Sulfur Activation in Hiroshima*

George D. Kerr and Joseph V. Pace III
Oak Ridge National Laboratory
Oak Ridge, Tennessee

We were asked to talk about a topic of special interest during this session. There are two special topics that I could have talked about, one being some microdosimetric calculations for active marrow in the skeleton. This work was done with Keith Eckerman and a preliminary report was published recently in the Proceedings of the Fifth Symposium on Neutron Dosimetry.[1] The other topic which I will discuss today deals with sulfur activation by fast neutrons from the Hiroshima bomb. Joe Pace was involved in this work from the very start, and he is listed as a co-author because I will be using some of his results in today's talk.

In 1979, we attempted to establish the validity of source terms for the Hiroshima and Nagasaki bombs using experimental data on sulfur activation.[2-4] Close agreement was observed between measured and calculated values for test firings of Nagasaki-type bombs.[4] The calculated values were based on source terms developed by W. E. Preeg at the Los Alamos National Laboratory (LANL).[5] A discrepancy was found, however, when we compared calculated values for the two bombs because a 1956 report by R. R. Wilson stated that sulfur activation by fast neutrons in Hiroshima was approximately three times greater than in Nagasaki.[6] Our calculations based on Preeg's source-term data predicted about equal sulfur activation in the two cities.

*Research sponsored by the Office of Health and Environmental Research, U.S. Department of Energy under contract DE-AC05-840R21400 with Martin Marietta Energy Systems, Inc.

Within a few weeks after the bombings, Japanese scientists measured the ^{32}P radioactivity produced by the $^{32}S(n,p)^{32}P$ reaction in sulfur which was used as a glue in electric power-line insulators. A detailed set of data on sulfur activation in Hiroshima was readily available from a 1953 report by Yamasaki and Sugimoto.[7] Their measurements of the beta particles from the decay of ^{32}P were made through a 0.0015-mm aluminum window of a calibrated Lauritsen electrometer. No equivalent data from measurements using a calibrated detector were found for sulfur activation in Nagasaki. Another 1953 report was found, however, which stated that sulfur activation at the hypocenter in Hiroshima was about 60% less than that observed at the hypocenter in Nagasaki. [8] With this statement, we were able in 1980 to resolve the apparent discrepancy in our earlier work.

As noted previously, the calculated values based on Preeg's source-term data predicted about equal sulfur activation in the two cities, but the calculated values were found to be greater than the measurements at the hypocenter in Hiroshima (see Figure 1). [4] This difference was consistent with the general observation that sulfur activation at the hypocenter in Hiroshima was less than that at the hypocenter in Nagasaki. [8] Thus, we suggested that the difference in sulfur activation at the hypocenters in the two cities was the result of a blind spot in the neutron leakage through the nose section of the Hiroshima bomb. The blind spot produced a highly asymmetric-neutron leakage about the major axis of the cylindrical gun-assembly device.[4, 9]

Simplified one-dimensional models of the bombs were used in the development of Preeg's source-term data. A one-dimensional model was adequate for the Nagasaki bomb because it was a spherical symmetric implosion device with a nearly isotropic neutron leakage. [5] The Hiroshima bomb, however, required the use of a two-dimensional model to obtain information on the asymmetry in the neuton leakage. Such a model was developed and used in the 1983 work by Whalen and his colleagues at LANL. [10] Their source-term data provided much closer agreement between calculated and measured values for sulfur activation at the hypocenter in Hiroshima (see Figure 2). [11] At large distances from the hypocenter, the agreement was not very good, but the uncertainties in the measured values beyond 500 m were quite large according to a 1983 review by Hamada. [12] His review also provided information on the physical characteristics of the porcelain insulators from a 1958 report by Yamasaki.[13]

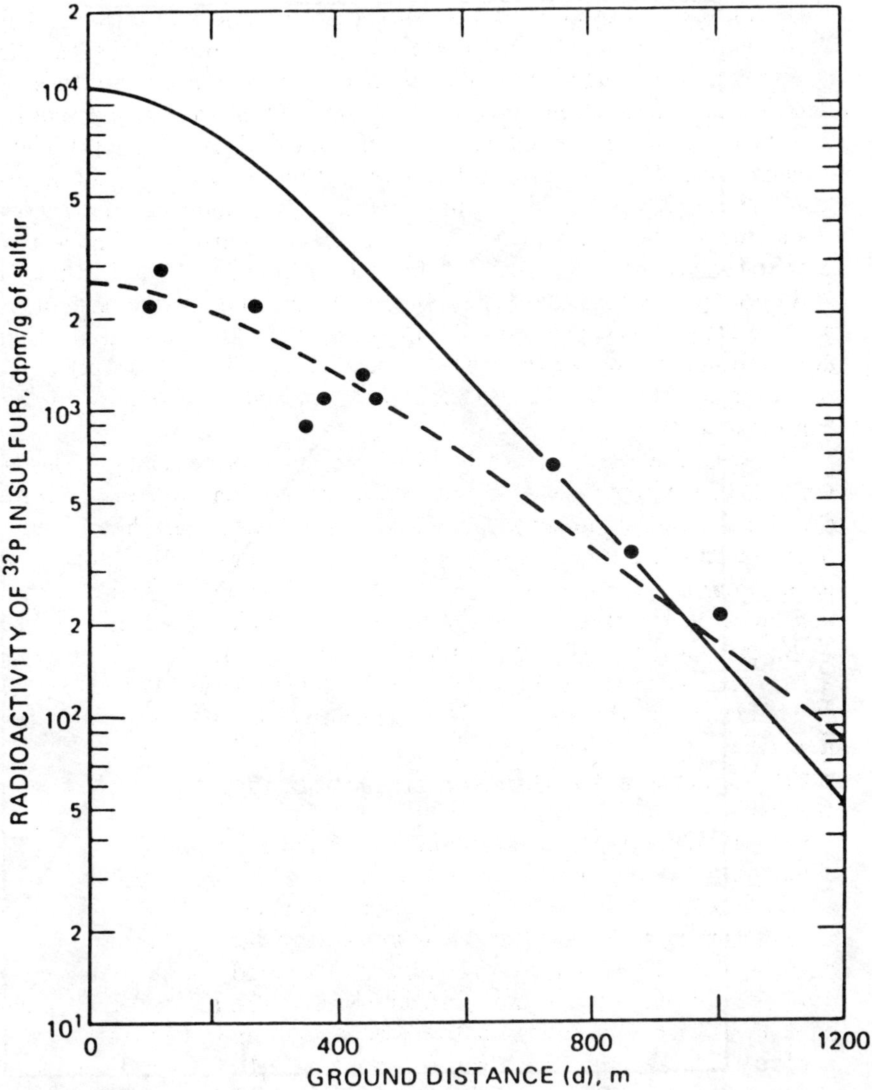

Fig. 1. Comparison of calculated values (solid line) and measured values (circles) for sulfur activation in Hiroshima. The calculated values are based on Preeg's source-term data and a bomb yield of 12.5 kilotons (References 2 and 4). The dashed line shows a least squares fit to the measured values from a 1953 report by Yamasaki and Sugimoto (Reference 7).

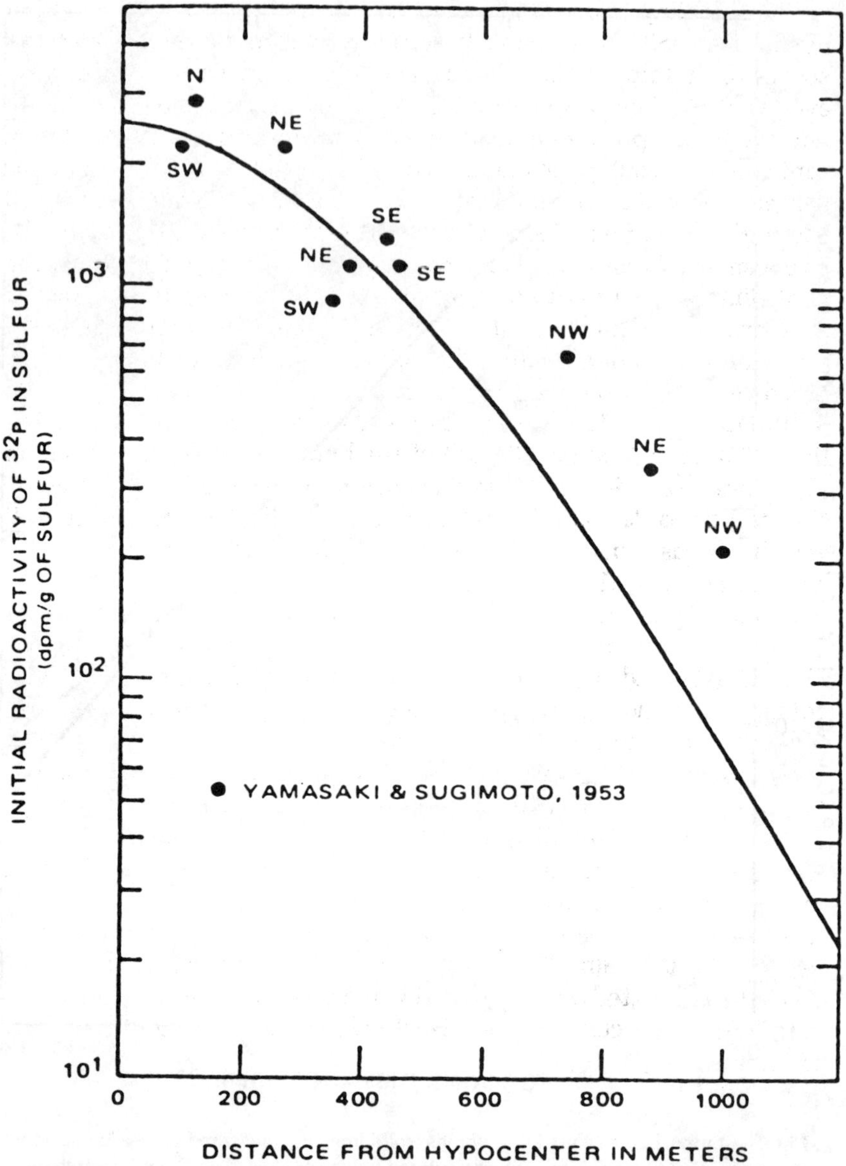

Fig. 2. Comparison of calculated values (solid line) and measured values (circles) for sulfur activation in Hiroshima. The calculated values are based on Whalen's source-term data and a bomb yield of 12.5 kilotons (Reference 11). The measured values and their directions relative to the hypocenter are taken from the 1953 report by Yamasaki and Sugimoto (Reference 7).

The shielding effect of the porcelain insulators was considered in our initial 1983 calculations of sulfur activation based on Whalen's source-term data for the Hiroshima bomb.[11] It was assumed, however, that the bomb was oriented vertically at the time of explosion and the blind spot was pointed directly downward toward the hypocenter. As a result, our calculated values tended to be too low for samples located at small distances to the east of the hypocenter and too high for samples located at small distances to the west of the hypocenter (Figure 2). We had previously determined that the exploding bomb was tilted about 15° and the blind spot was pointed at a ground location about 150 m from the hypocenter along the direction of bomber heading (i.e., almost due west or 265° as measured clockwise from true north). Thus, a more elaborate calculation was made in late 1983 to assess the effects of the bomber heading (BH 265) and tilt angle (TA 15) of the bomb on sulfur activation in Hiroshima (see Figure 3).[14] The calculated results broadly showed that the tilt of the bomb (BH 265, TA 15) affected the induced ^{32}P activity by as much as 20% when compared to our calculations for an untilted bomb (BH 270, TA 0).

Finally, the calculated results were compared with revised values for sulfur activation in Hiroshima from the 1983 review by Hamada[12] and 1958 report by Yamasaki.[13] In making these comparisons and others, we assumed a bomb yield of 12.5 kilotons (Figures 1 through 3). It is possible, however, to derive a weighted mean value for the bomb yield using Hamada's newer sulfur-activation data.[12, 15] These data and our calculations for a tilted bomb (BH 265, TA 15) predict a bomb yield of approximately 13 ± 2 kilotons. About one half of the standard deviation of 2 kilotons arises from uncertainties in the measured values and the other half arises from uncertainties in the calculated values due to the cross sections for the $^{32}S(n,p)^{32}P$ reaction. The weighted mean value derived from the calculated and measured values for sulfur activation agrees closely with the currently accepted value of 15 ± 3 kilotons for the Hiroshima bomb.

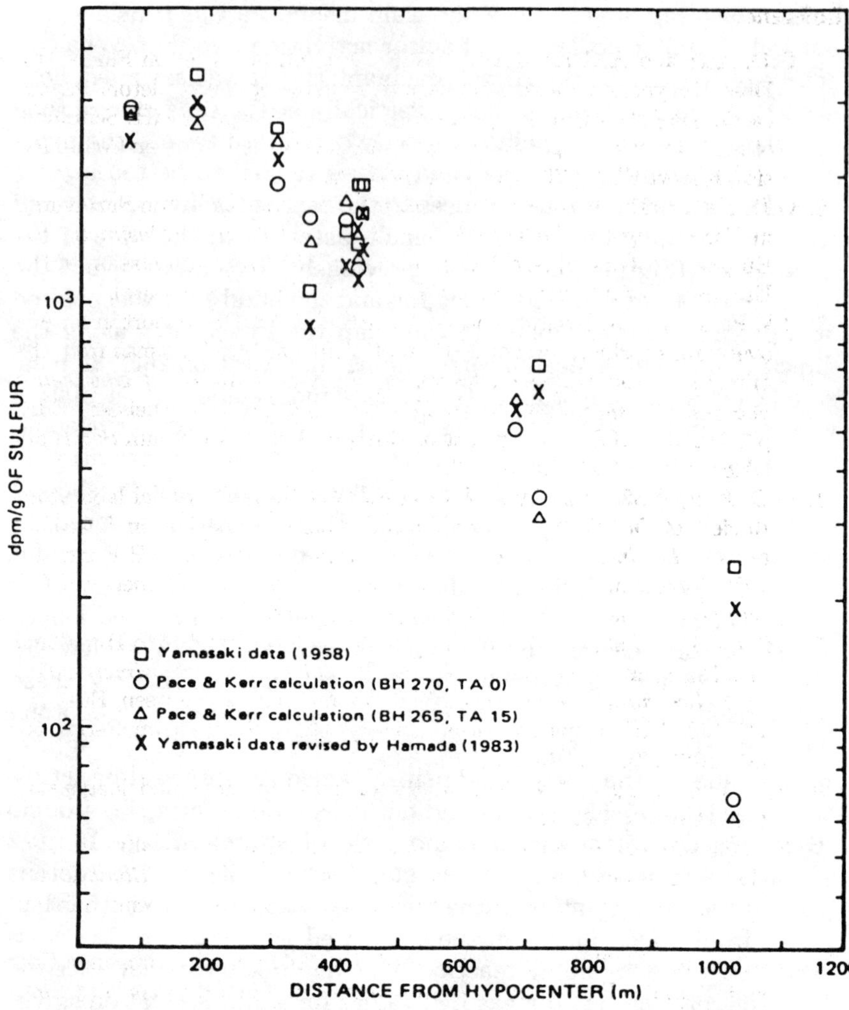

Fig. 3. Comparison of calculated and measured values for sulfur activation in Hiroshima. The effect of bomber heading (BH) and tilt angle (TA) of the bomb were considered in these calculations which are based on Whalen's source-term data and a bomb yield of 12.5 kilotons (Reference 14). The measured values are revisions to the original sulfur-activation data from a 1958 report by Yamasaki (Reference 13) and a 1983 review by Hamada (References 12 and 15).

References

1. G.D. Kerr and K.F. Eckerman, 1985, "Neutron and Photon Fluence-to-Dose Conversion Factors for Active Marrow of the Skeleton, in *Proceedings of the Fifth Symposium on Neutron Dosimetry* (H. Schraube, et al., Eds.) Vol. 1, pp. 133-145 (Commission of the European Communities, Report EUR 9762 EN, Luxembourg).
2. G.D. Kerr, 1981, "Review of Dosimetry for the Atomic Bomb Survivors," in *Proceedings of the Fourth Symposium on Neutron Dosimetry* (G. Burger and H.G. Ebert, Eds.) Vol. 1, pp. 501-513 (Commission of the European Communities, Report EUR 7448 EN, Luxembourg).
3. J.V. Pace III, J.R. Knight, and D.E. Bartine, 1982, "Transport in an Air-over-Ground Environment of Prompt Neutrons and Gammas from the Hiroshima and Nagasaki Weapons," in *Reevaluations of Dosimetric Factors: Hiroshima and Nagasaki* (V.P. Bond and J.W. Theissen, Eds.) pp. 131-158 (U.S. Department of Energy, DOE Symposium Series 55, DE81026279, Oak Ridge, TN).
4. G.D. Kerr, 1982, "Findings of a Recent Oak Ridge National laboratory Review of Dosimetry for the Atomic Bomb Survivors," in *Reevaluations of Dosimetric Factors: Hiroshima and Nagasaki* (V.P. Bond and J.W. Thiessen, Eds.) pp. 52-97 (U.S. Department of Energy, DOE Symposium Series 55, DE81026279, Oak Ridge, TN).
5. P.P. Whalen, 1982, "Status of Los Alamos Efforts Related to Hiroshima and Nagasaki Dose Estimates," in *Reevaluations of Dosimetric Factors: Hiroshima and Nagasaki* (V.P. Bond and J.W. Thiessen, Eds.) pp. 111-130 (U.S. Department of Energy, DOE Symposium Series 55, DE81026279, Oak Ridge, TN).
6. R.R. Wilson, 1956, "Nuclear Radiation at Hiroshima and Nagasaki," *Radiat. Res. 4*, 349-359.
7. F. Yamasaki and A. Sugimoto, 1953, "Radioactive ^{32}P in Sulfur in Hiroshima," in *Collection of Investigative Reports on Atomic Bomb Disaster*, Vol. 1., pp. 16-18 (Japanese Science Promotion Society, Tokyo, Japan).
8. T. Miyazaki and T. Masuda, 1953, "A-Bomb Radiation in Hiroshima City and Vicinity - Part 2," in *Collection of Investigative Reports on Atomic Bomb Disaster*, Vol. 1., pp. 35-37 (Japanese Science Promotion Society, Tokyo).
9. G.D. Kerr, J.F. Emery, and J.V. Pace III, 1985, *Sulfur activation at the Little Boy-Comet Critical Assembly: A Replica of the Hiroshima Bomb*, Oak Ridge National Laboratory Report ORNL/TM-9439.
10. P.P. Whalen, 1983, "Source Terms for the Initial Radiations," in *Reassessment of Atomic Bomb Radiation Dosimetry in Hiroshima and Nagasaki*, pp. 13-44, Proceedings of a joint U.S.-Japan workshop held at Nagasaki, Japan, February 16-17, 1983 (Radiation Effects Research Foundation, Hiroshima).
11. G.D. Kerr, J.V. Pace III, and W.H. Scott Jr., 1983, "Tissue Kerma vs. Distance Relationships for Initial Nuclear Radiation from the Atomic

Bombs," in *Reassessment of Atomic Bomb Radiation Dosimetry in Hiroshima and Nagasaki*, pp. 57-103, Proceedings of a joint U.S.-Japan workshop held at Nagasaki, Japan, February 16-17, 1983 (Radiation Effects Research Foundation, Hiroshima).

12. T. Hamada, 1983, "Measurement of ^{32}P Activity Induced in Sulfur in Hiroshima," in *Reassessment of Atomic Bomb Radiation Dosimetry in Hiroshima and Nagasaki*, pp. 45-56, Proceedings of a joint U.S. -Japan workshop held at Nagasaki, Japan, February 16-17, 1983 (Radiation Effects Research Foundation, Hiroshima).
13. F. Yamasaki, 1958, "Estimation of Radiation Dose from the Bombing Attack," in *Annual Report of Cooperative Research, Japanese Ministry of Education*, Radiation Section 33, p. 121.
14. J.V. Pace III and G.D. Kerr, 1984, "Sulfur Activation in Electric Insulators in Hiroshima," in *Reassessment of Atomic Bomb Radiation Dosimetry in Hiroshima and Nagasaki*, pp. 56-58, Proceedings of a joint U.S.-Japan workshop held at Hiroshima, Japan, November 8-9, 1983 (Radiation Effects Research Foundation, Hiroshima).
15. T. Hamada, 1984, "^{32}P Activity Induced in Sulfur in Hiroshima: Reevaluation of Data by Yamasaki and Sugimoto," in *Reassessment of Atomic Bomb Radiation Dosimetry in Hiroshima and Nagasaki*, pp. 52-55, Proceedings of a joint U.S.-Japan workshop held at Hiroshima, Japan, November 8-9, 1983 (Radiation Effects Research Foundation, Hiroshima).

Perspectives on Radiation Dose Estimates for A-Bomb Survivors

William E. Loewe
Lawrence Livermore National Laboratory*,[1]

Abstract

Four decades after the actual events, quantitative characterization of the radiation fields at Hiroshima and Nagasaki continues to be sought, with high accuracy a goal justified by the unique contribution to radiation protection standards that is represented by the medical records of exposed survivors. The most recent effort is distinguished

---

*This work was performed under the auspices of the U.S. Department of Energy by Lawrence Livermore National Laboratory under contract No. W-7405-Eng-48.

by its reliance on computer modelling and concomitant detail, and by its decentralized direction, both internationally and internally to the U.S. and Japan, with resultant ongoing peer review and wide scope of inquiry.

A new system for individual dose estimation has been agreed upon, and its scientific basis has been elaborated in the literature as well as in a comprehensive treatise to be published in the Spring of 1987. In perspective, this new system appears to be an unusually successful achievement that offers the expectation of reliable estimates with the desired accuracy. Some aspects leading to this expectation, along with a *caveat*, are discussed here.

Introduction

The need for estimating radiation dose to specific organs of individuals surviving the Hiroshima and Nagasaki bombings may represent a unique scientific enterprise, in the following sense. There is a strong global public significance attached to absolute accuracy in these dose assignments at the level of a few tens of percent accuracy, for radiations that occurred in a non-reproducible event of high physical complexity, four decades ago under the extraordinary conditions of war and with little opportunity for contemporary observation. The public significance derives from the fact that the medical records of exposed survivors represent the only body of data relevant to radiation risk evaluation that applies to effects of whole-body radiation on humans, which includes a relatively large statistical sample representing a general population, and for which record details are available on a generally consistent basis.

This challenge to the physics community has been met over the years with various approaches. One is to use direct analogy from observation of other nuclear explosions, such as the Trinity event in New Mexico and the Ranger Fox event in Nevada for comparison to the radiatively analogous explosion of the "Fat Man" bomb at Nagasaki. (No analogous observed event exists for the explosion of "Little Boy" at Hiroshima.) Another approach is to use the phenomenology of radiation in related situations, such as suspended nuclear reactors, a procedure comprising the primary effort on the previous dosimetry system called T65D ("tentative doses in 1965"). A third approach is to infer radiation levels from subsequent measurements of radiation-affected materials at the actual sites themselves, as has been done by Japanese workers using activated sulfur and cobalt for neutron estimates and using thermoluminescence techniques for gamma estimates. A fourth approach is to analyze and model quantitatively the physical mechanisms that determine radiation levels, at least to the degree required to insure that inferences and analogies are not misapplied and that assumptions are reasonable.

The recently culminated, half-decade program of joint effort by Japanese and American scientists that has lead to the "DS86" ("dosimetry system as of 1986") set of dose assignments has included the work described in the previous papers of this session. The program was strongly shaped by analysis of physical mechanisms, especially as modern computer capabilities enable detailed investigation, but in fact the program exploited all of the previous approaches, with strong reliance on observation of analogous events for quantitative model validation, and with emphasis on *in-situ* data, extended and reinterpreted as part of the program, for evaluation of the computer predictions. It is this comprehensive synthesis of approaches, ordered by analysis of physical mechanisms and carried out with the detail made possible by computers, that provides one of the major reasons for investing the results with a high degree of confidence in their essential reliability. An early expression of this kind of synthesis appeared in the archival literature as References 1 and 2. The full expression of this overall approach appears in a comprehensive treatise describing all details of the program, scheduled for publication in Spring of 1987 by the Radiation Effects Research Foundation, headquartered in Hiroshima. Oversight for the joint program was provided by Dr. Eizo Tajima of Tokyo and by Dr. Frederick Seitz of New York, with Dr. Robert F. Christy of Pasadena as Group Leader for the U.S. effort.

Discussion

During the first two decades since the events at Hiroshima and Nagasaki, various estimates based on occasional expert opinion showed revisions that in some cases amounted to an order of magnitude. In 1965, a substantial several-year program, centered on measurements of reactor radiations at the Nevada Test Site (NTS), produced a dose assignment system called T65D which was accepted until 1980, when new estimates were set forth formally and publicly which revised assignments by factors of as much as five to ten. The immediate impact was to remove the Hiroshima and Nagasaki data as sources of information on neutron risk. Since 1980, estimates of the radiation field have changed very little, as shown in Figs. 1 and 2, which give (a) free

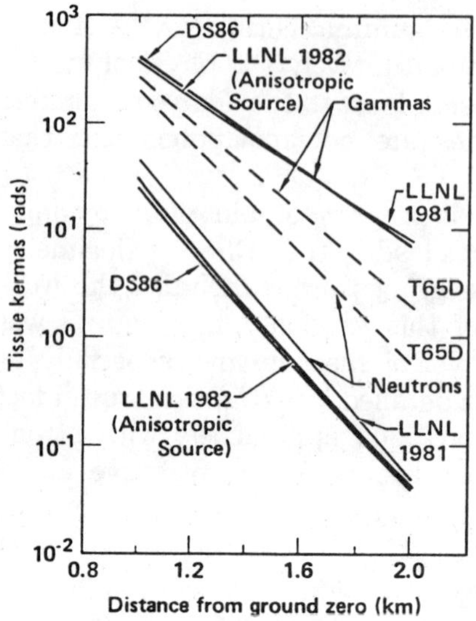

Fig. 1. Free-in-air tissue kermas at Hiroshima

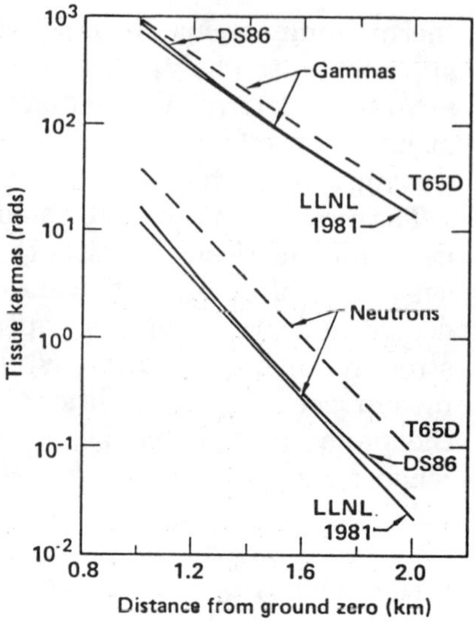

Fig. 2. Free-in-air tissue kermas at Nagasaki

field kermas obtained as the T65D results, (b) the 1980 results from Reference 3 with subsequent adjustment in 1982 for anisotropic sources (cf. Paul Whalen in this session), and (c) the DS86 results. Similarly, Table 1 shows the average house shielding factors for neutrons and gammas at Hiroshima, obtained as the T65D values, the 1980 values given by Jess Marcum of RDA, and DS86 values (cf. W.A. Woolson *et al.*, in this session). Comparable results are obtained for Nagasaki.

The remarkable feature of Figures 1 and 2 and Table 1, another major reason for high confidence in reliability of DS86 dose assignments, is the small changes brought about by the scrutiny after 1980, in which all aspects of all calculations were repeated and frequently improved, usually in both countries, and validating or supporting observations were carefully re-evaluated and frequently extended. This suggests that methodologies have matured, and that results are not overly sensitive to reasonable assumptions about meteorological conditions, so that no large changes are to be expected in the future. In fact, future changes larger than perhaps 10% at a time (and very few of them) are presently regarded by the community as being quite unlikely.

TABLE 1—*Average Transmission Factors at Hiroshima*

Source	Radiation		
	Neutron	Prompt Gamma	Delayed Gamma
T65D	0.32	0.90	0.90
J. Marcum (1980)	—	0.55	0.45
DS86	0.38	0.53	0.46

Some special features of the revisions introduced in 1980 were (a) explicit use of meteorology particular to the events, made possible in part by reliance on calculational models validated elsewhere that permitted direct extrapolation to other meteorological conditions, and (b) the use of separate components, with models individually validated by experimental data, for gamma dose estimates. Although well known by specialists to be otherwise, the interested physics community believed in 1980 that the T65D modelling of gamma dose, in which attenuation goes as $\exp(-R/\lambda)/4\pi R^2$, was correct after a short transient distance. In fact, the two major components (coming mostly from neutron capture in atmospheric nitrogen on the one hand, and from fission product debris on the other hand), each have quite different e-folding lengths but similar magnitudes, and cannot add up to the T65D form.

An additional special feature was use of detailed radiation leakage intensities as a function of energy for each bomb, made possible by ever-more-detailed calculations at Los Alamos and discussed in this session by Paul Whalen. Figure 3 shows Little Boy spectra for neu-

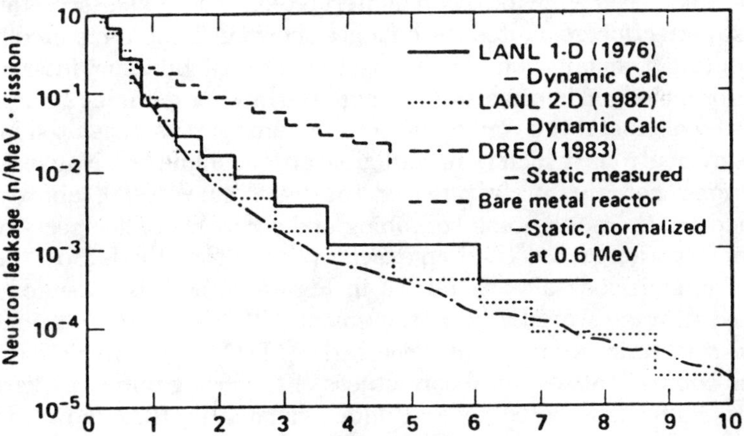

Fig. 3. Little Boy neutron leakage energy spectra

trons with energy great enough to penetrate the atmosphere to distances of greatest interest to survivor application. (See Table 2 as an example of how higher source energies increase in importance with distance.) The bare reactor data represent the basis for the T65D neutron estimates (actually obtained phenomenologically at a reactor), while the remaining curves compare calculation and experiment at the most copiously emitting angle for the static replica as well as the exploding bomb. The 4–5 tons of heavy metal (iron and tungsten) in the gun assembly imply *a priori* that these data must be roughly right for the bomb, and the replica results confirm accuracy. The small change shown from static to dynamic is to be expected from the fact that first-principle estimates based on reactor physics and sound speeds show that neutrons at these energies are generated and exit well before any major changes in the overall geometry.

TABLE 2—*Contribution to Kerma of Source Neutrons ≥3 MeV*

Range (km)	Fraction of Total Kerma
1.0	30%
1.5	50%
2.0	70%

(Only neutrons with energies ≥ 3 MeV can activate sulfur.)

Other 1980 features were (a) extensive reliance on geometrically correct and validated radiation transport calculations, discussed in Ref. 2 and partially abstracted with updates in Table 3, and (b) recognition by Jess Marcum of RDA that measured building gamma-ray transmissions had been previously misinterpreted. The misinterpretation, whose correction generated the changes shown in Table 1, resulted from computing transmission factors as the ratio of building interior to exterior measured kermas, even though the interior gamma-ray kermas included a substantial contribution from gamma rays generated in the walls by neutrons (which were negligible at Hiroshima and Nagasaki).

Among noteworthy features of the work since 1980, above and beyond the thorough-going scrubbing and scrutiny of all aspects, were (a) construction at LANL of an exact replica of the "Little Boy" bomb (spectral emissions already shown in Figure 3), and (b) extended and revised estimates of *in-situ* measurements. These *in-situ* data include (a) thermoluminescence data discussed by D.C. Kaul in this session, which confirm the essential correctness of the new gamma-ray kermas out to half a kilometer beyond data available in 1980 (now to 1½ kilometers from the hypocenter at Hiroshima), and (b) recent data on europium activation, as well as (c) new detailed comparisons between measured and calculated sulfur and cobalt activation. The europium and sulfur comparisons are discussed by M. Gritzner *et al.*, in this session, with bottom lines (a) that data-spread for europium prevents either confirmation or contradiction of calculated low energy neutrons, and (b) that sulfur shows excellent agreement at short distances, implying generally good absolute intensities of neutrons that will subsequently penetrate to large distances of interest (note Table 2 again), but at distances approaching a kilometer, error bars are too large to permit conclusions about calculated high energy neutron intensities.

Comparisons of calculated and measured cobalt activations, however, thought in 1980 to be in agreement on the basis of simplified calculations, now clearly show major contradictions as depicted in Figure 4 from Ref. 4. On the one hand, all other comparisons and considerations (including several discussed above) support calculated estimates of higher energy neutrons, and the correctness of then calculating lower energy neutrons is supported by knowledge of physical mechanisms involved as well as by comparison with low energy neutron measurements at the Ranger Fox event (similar to Nagasaki) and at the APRD reactor (which generated the data shown in Table 3). On the other hand, these cobalt measurements have been reviewed carefully, and new data obtained at other locations in Hiroshima are entirely consistent with the old.

Other special features of the work since 1980 include (a) continuing review of explosive energy estimates and benchmarking with the Little Boy replica (no change in yield for Hiroshima and 5% change for Nagasaki from Ref. 3), and (b) concerted effort to establish bounds on survivor dose from fallout and induced activation, resulting in upper bound estimates for those receiving significant contributions. For that small fraction of the survivors spending much time in those limited areas involved, absorbed doses due to fallout and induced activation are significant but modest.

TABLE 3—*Transport Calculations for APRD*

Range (m)	Calculated/Measured	
	Neutron kerma	Gamma kerma
100	1.08[a]	1.06
170	1.06[a]	0.91
300	1.30[a]	0.94
400	1.28[a]	0.90
1,080	0.92[a]	0.80
1,618	0.90	0.91

[a]Revised 1984

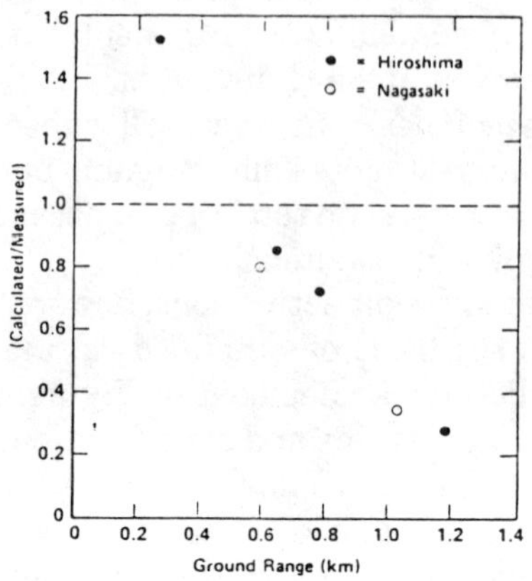

Fig. 4. Cobalt activation

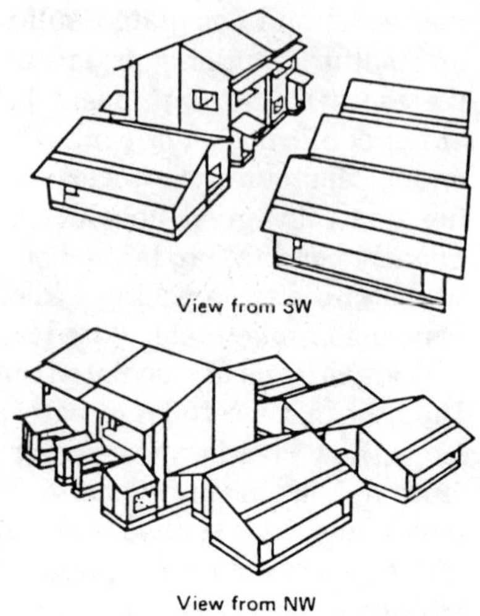

Fig. 5. Six-house cluster

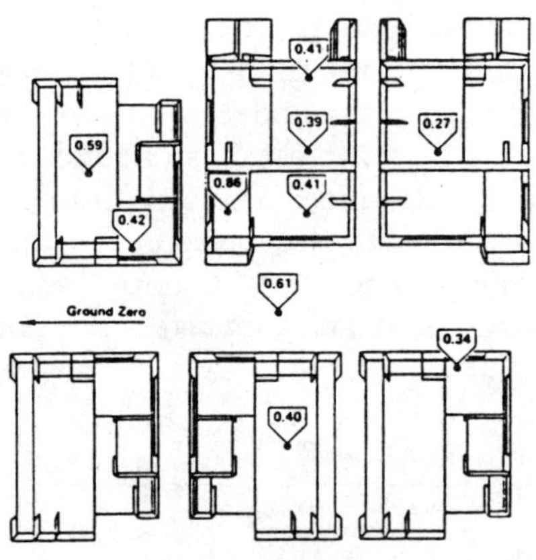

Fig. 6. Prompt gamma ray transmission factors

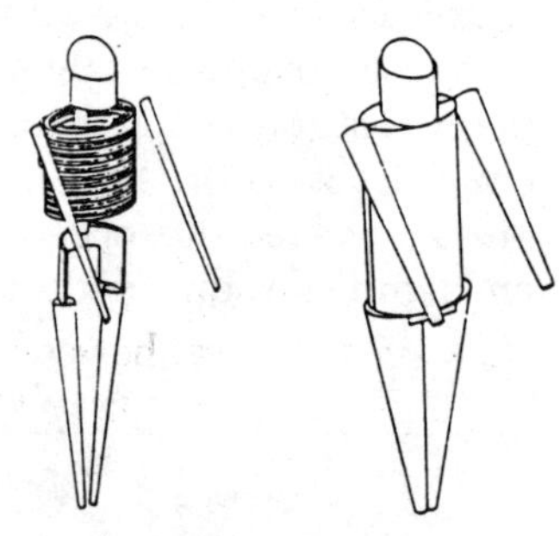

Fig. 7. Adult phantom and skeleton

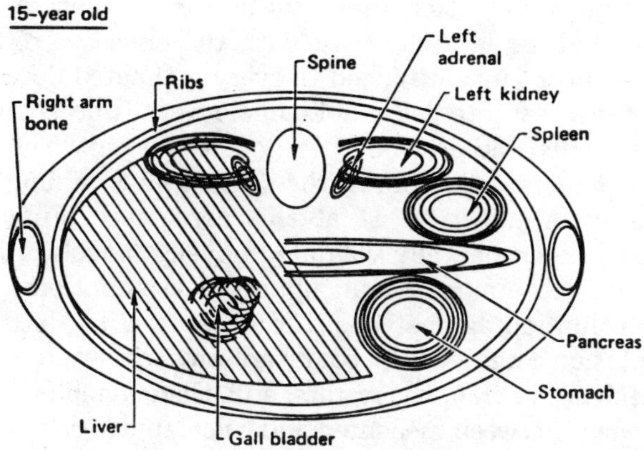

Fig. 8. Cross-section of trunk

Probably the most important and certainly the biggest effort since 1980 has been the careful development of detailed, fully consistent and accurate estimates of the shielding provided by both surrounding buildings and by the body itself, as previously discussed at this session by W.A. Woolson *et al.* and by Steve Egbert of SAI. Based on a long-available 3D method developed by ORNL and validated by SAI, with further validation since 1980 to better than five percent when compared to data from full-scale Nevada mockups of Japanese houses, these estimates result from detailed representation of both geometry and angle-position-energy-dependent physical mechanisms. They give unprecedented realism to estimates of neutron or gamma ray dose to a particular organ of a particular individual located in a particular environment in a specific place, including all known physical contributions. Figures 5 through 8 illustrate the geometrical complexity actually used. Accuracy in these shielding factors, which varies by as much as a factor of three, is believed to be determined by record keeping, not by the estimation process.

Conclusions

The current status is that a very complete report supporting the formally accepted DS86 results was prepared in 1986, and a system was installed at Hiroshima for obtaining specific organ doses for particular individuals, which includes the capability of readily accepting any remaining adjustments that may still be forthcoming. A substantial uncertainty analysis is under way, with the objective of providing a quantitative uncertainty attached to every estimated dose; for typical survivors of greatest interest to epidemiology, the uncertainties in dose assignment are expected to be a few tens of percent. This analysis is discussed in papers appearing in the session on Shielding and Radiation Transport Application, II at this conference. Fully consistent application of experimentally validated physics models of all radiation components, with almost no unsupported assumptions, along with extensive evaluation against *in-situ* data as well as data from other nuclear explosions, has lead the cognizant physics community to invest strong confidence in all of the results; still, there remains one outstanding discrepancy, between measured and calculated *in-situ* cobalt activation (by thermal and slightly epithermal neutrons). In progress are attempts to (1) reduce data scatter in *in-situ* europium activation measurements and to (2) resolve differences of expert opinion on the suitability of zircon and uranium-bearing glass samples for fission track analysis; both are independent measures of neutrons similar to those that activate cobalt.

In 1981, intense international interest in the dose revisions was usually expressed to me as the question, "What impact does this have on our understanding of radiation risk," along with expressions of lost innocence in finding that we physicists had previously been so wrong. I like to think that we have now regained much of that lost confidence, and that just maybe the impact will appear first as a shift of some epidemiological data from the class of Rorschach ink blots to the class of scientific evidence, from which new insights may appear.

References

1. W.E. Loewe and E. Mendelsohn, "Neutron and Gamma-Ray Doses at Hiroshima and Nagasaki," Nuc. Sci. Eng. *81*, 325–350 (1982).
2. W.E. Loewe, W.A. Turin, C.W. Pollock, A.C. Springer, and B.L. Richardson, "Validated Deep-Penetration, Air-Over-Ground, Neutron/Gamma-Ray Transport," Nucl. Sci. Eng. *85*, 87–115 (1983).
3. W.E. Loewe and E. Mendelsohn, "Revised Dose Estimates at Hiroshima and Nagasaki," Health Phys. *41*, 663–666 (1981).
4. W.E. Loewe, "Hiroshima and Nagasaki Initial Radiations: Delayed Neutron Contributions, and Comparison of Calculated and Measured Cobalt Activation," Nuc. Tech. *68*, 311–318 (1984).

Source Spectrum and Output Spectrum Calculations

Paul P. Whalen
Los Alamos, New Mexico

My part of this program has been very small, just associated with the source terms. I didn't even do any of the work associated with the source terms, it was all done by various people in our laboratory who performed the calculations; all more expert at it because they had had years of practice which I hadn't had. But I want to start by addressing a subject left from this morning. There were two comments, one about the induced activity measured on the ground, and the other about the Alvarez experiments. John Harley covered the one, indeed that data was used. I want to reassure everyone that it was not overlooked. The data are all in John Harley's chapter. And the Alvarez data are in John Malik's chapter on the yield of the Hiroshima explosion. When I finish my talk, I want someone to ask John Harley a question about rainfall in Japan, particularly in 1945. John Auxier, in the Symposium in 1981, addressed a question of: "O.K., if we do all these calculations, will they mean anything, or will they be just . . . calculations?" The laboratory with which I am associated went to a great deal of trouble trying to calibrate the calculations of the source term by putting together the replica, of which you have seen the pictures, and inviting lots of people to do measurements. Now, if you recall those curves, the scatter in the data looks like ± 15 to 20 percent. That is scatter in the measurements, not in the calculations. This will be talked about again in a couple of weeks in the meeting which Bill Loewe mentioned that has to do with measuring radiation. But, for Auxier's benefit, the spectra measured on a spherical replica of the little-boy device built in 1965, and reported by Thorngate *et al.* were a real thorn in our side, because they did not look like our calculations. After having had some 10 or 12 different experimenters do measurements on the new Little-Boy mockup, they do look like (to the ± 20 percent discrepancy) our calculations and not like the Ichiban measurements. This was extremely satisfying to us, and one of the reasons why I have not changed the calculations although they indeed could stand to be changed. There are two reasons for redoing the output calculations: The ENDF B Mod 4 corrections to the iron cross sections which were referred to this morning, are indeed in the right direction. I have done calculations and made comparisons with the measurements (they will be in the final report of this working group). They reduce, in the high energy limit, the sulphur activation and the neutrons above 1 MeV. They reduce the scatter in the comparison with the experiments to about 15 percent from the 20 percent that you saw in the curves this morning. Actually, if you look at it in great detail, they overcorrect points which were too high and all of a sudden become too low, reducing the spread between the different measurements. Another reason for redoing the output calculations, would be to report the results in a more detailed group structure. As a matter of fact, since the explosion calculation has been done and saved on what we call mass storage, it would not be all that difficult. It appears, in some interesting work that SAI is doing now, that, if group boundaries were chosen more appropriately so that they coincided with windows and lack of windows in air cross sections they could get some small change in the agreement at distance, due to the air transport.

I've been asked to address this question of time scales. Could the Hiroshima bomb have cracked open during the explosion? The answer is no, there is no way in the world that the Little-Boy bomb could have cracked open during the explosion. I guess the way to talk about this is to consider thunder and lightning. The speed of sound in air is about 300 m/s so if you see lightning and count to five before you hear thunder, that indicates the lightning struck a mile away. Well, the speed of sound in steel is quite a bit higher, about 10 times higher, and so that indicates a few mm/μs. The explosion is over in less than a μs; so any rarefaction or compression wave, which starts out from the explosion associated with this thing coming apart, can travel only a few mm. The working part of that device was machined out of a piece of steel 36 in. long, with nothing in there that will move more than a few inches in the entire time of the explosion (except for the explosion itself which is moving very high speed shock waves and disassembling the fissile material). I don't know how to address it much better. No, the case did not crack apart. It was impossible.

One of the items which hasn't been covered very strongly here, is the yield of the device. John Malik put together the chapter which covered all of the different measurements, error estimates and so forth, and came up with the yield of 15 kT. I don't want to discuss that, but in 1981 I said that we would build the replica of the device, measure the critical separation (the point at which we bring the uranium into the bomb) and make an estimate of the maximum yield. I can only estimate the maximum yield, because if something goes wrong, it can fizzle and give no yield, or a very small yield. That program was done. Exact replicas of the parts of the fissile materials were cast, and the experiment was performed. Dean Kaul referred to it this morning. The measured critical separation distance and the one that I calculated, agreed to 0.01 cm. Having found great agreement on the critical separation, I then calculationally moved the uranium totally into the device, calculated its reactivity and looked at how it compares with what we used for the explosion calculation. We were right on, our calculated yield was 15 kT. The range which we could have covered when the explosion initiated was 13 to 16 kT. Actually, on the basis of changing cross sections, different shock wave patterns and so forth, I raised that to an upper limit of 17 kT. This estimate includes the work done by our sister laboratory, Livermore, on the yield to be expected *ab initio* from that calculation. I throw in 10 percent because we have no measurements of yield which we can qualify as better than ± 10 percent so the absolute, upper limitation on the yield is then 19 kT. This only rules out 2 or 3 estimates which were higher than that. It's unfortunate. It would have been nice if it had resulted in a maximum yield of 12 kT, then we could settle on the old number, but it didn't come out that way. The actual yield estimated from the critical separation measurements is 15 kT, which was where we started out several years ago. The experiments which we did at Los Alamos were a real combination of factors, the fact that we had found the pieces of the device, the fact that the nuclear emergency search teams were very interested in calibrating detectors between various laboratories, and the fact that one organization was interested in promoting their facility. This whole combination of things just came together at a very appropriate time so we were able to do those experiments and really demonstrate that our neutron source calculations were quite good.

Scientific Session

The New Dosimetry and Its Application

Edward L. Alpen, *Chairman*

The Dosimetry System 1986 (DS86)

William A. Woolson
Science Applications International Corporation
San Diego, California

In May 1983, we proposed a dosimetry system for use by the Radiation Effects Research Foundation (RERF) that would incorporate the new calculations of the joint United States-Japan research on the reassessment of the A-bomb dosimetry. The dosimetry system features in the proposal were based on considerations of the output data required from the dosimetry system for the normal work of RERF and for future research in radiobiology; the computerized input data available for A-bomb survivors in the RERF data base; the level of detail, precision, and accuracy of various components of the dosimetric estimates; and the computer resources available at the RERF facility in Hiroshima.

The proposal evolved from the extensive discussions with RERF personnel, numerous meetings of the physicists and scientists from Japan and the U.S. involved in the dosimetry reassessment research, and requirements expressed by epidemiologists and radiobiologists on the various review panels. These discussions indicated that, in light of the expansion of computer and radiation technology and the desire for more detail in the dosimetry, an entirely new approach to the dosimetry system was appropriate. This resulted in a complete replacement of the former dosimetry system (T65D), as distinguished from a simpler approach involving a renormalization of T65D parameters to reflect the new dosimetry.

The fourth joint U.S.-Japan Dosimetry Workshop held in Hiroshima on March 16 and 17, 1986, reviewed the results and findings of the research to assess the A-bomb dose estimates and their incorporation into DS86. As a result, the U.S.-Japan atomic bomb radiation dosimetry committees formally approved replacement of T65D with DS86 for use by RERF for computation of doses to A-bomb survivors. Then in June 1986, the RERF Board of Directors officially adopted DS86 for all RERF studies.

The previous dosimetry used by RERF called T65D (for tentative 1965 dose) was based entirely on experimental data obtained from nuclear weapon test explosions at the Nevada Test Site, from the Operation BREN experiment, and from fixed source shielding experiments. [1] The T65D used a simple two-parameter formula to calculate the Free-in-Air (FIA) or free-field neutron and total gamma-ray kerma as a function of slant range from the hypocenter. Shielding was treated by multiplying the FIA kerma by a radiation transmission factor whose value depended on the survivor situation. At a later date, an additional transmission factor was applied to account for body shielding if organ specific doses were desired. The reliance on experimental data which could not replicate the conditions of the A-bomb explosions over Hiroshima and Nagasaki was the principal cause of the now well-documented errors in the T65D dosimetry [2].

In contrast to T65D which relied on experimental data, the DS86 was produced entirely from computer based calculational models employing Monte Carlo and numerical methods. These calculational procedures are capable of simulating the specific conditions of the A-bomb explosions. The overall DS86 model involves the integration of several separate stages in the dosimetry, each individually calculated; these are: the radiation source emitted by the A-bomb detonation, the FIA transport of the prompt and delayed radiation components, the shielding by houses, terrain, and factory buildings, and the attenuation of the radiation through the body to the organ site.

The methods and codes used to calculate these components shown in Table 1 represent the current technical state-of-the-art procedure for each application. They have been individually validated to the extent feasible by comparison with experimental data. Research in each technical area, e.g., FIA prompt and delayed radiation transport, house and terrain shielding, and body shielding, focused on producing the best validated calculations with identifiable precision for each component [2].

Since DS86 was to be constructed from computer models and data, there was an apparent broad range in the level of detail in the dosimetry that the system could furnish. For example, a simple renormalization and scaling of the T65D parameters could have been implemented. On the other hand, a state-of-the-art calculation could have been, in theory, performed for each survivor in the data base.

TABLE 1—*DS86 Computational Methods and Experimental Validation.*

DS86 Data Base	Calculation Method	Validating Experiments
Free-Field		
— Source	Coupled Radiation Hydrodynamics	Comet Assembly (Little Boy Replica) Sulfur Activation at Hiroshima
— Prompt	Two-Dimensional Discrete Ordinates	APRD Measurements; Weapon Test Data TLD Dosimetry at Hiroshima and Nagasaki
— Delayed	Phenomenological Fireball and Shock Hydrodynamics	Weapon Test Data TLD Dosimetry at Hiroshima and Nagasaki
	One-Dimensional Radiation Transport with Ground Corrections	
House Shielding	Adjoint Monte Carlo	Bren House Shielding Experiments TLD Dosimetry at Nagasaki
Body Shielding	Adjoint Monte Carlo	Laboratory Experiments

The first option, however, would not realize the full sophistication of the new dosimetry nor meet the data and accuracy requirements for RERF. The latter would not be feasible for the RERF computer facility, nor cost-effective.

The selected architecture for DS86 uses stored data bases saved from detailed computations for each stage of the dose estimate. These data bases maintain the full differential level of the original calculations. Each data base contains many separate data sets, corresponding to a complete description for different configurations or cases, such as ground range in the free field data base, type of house and location within the house for the house shielding data base, and organ site, posture, etc., in the body shielding data base. As depicted in Figure 1, given the available computerized information for a survivor in the RERF data files, the system selects the most appropriate elements from each data base and combines them to give a fully detailed representation of the radiation environment for the free field at the survivor's ground range, for the shielded exposure site, and for each of 15 organs of the body.

The radiation environment is computed as energy and angular-dependent fluence resolved into 58 energy groups (37 neutron, 21 gamma-ray) and 240 solid angle bins. DS86 reduces the energy-angle differential fluence data to a set of kermas and energy dependent fluence for output. The energy dependent fluence (or fluence spectrum) is simply an integral over solid angle of the differential fluence:

$$\phi(E) = \int_{4\pi} \phi(E, \Omega)d\Omega,$$

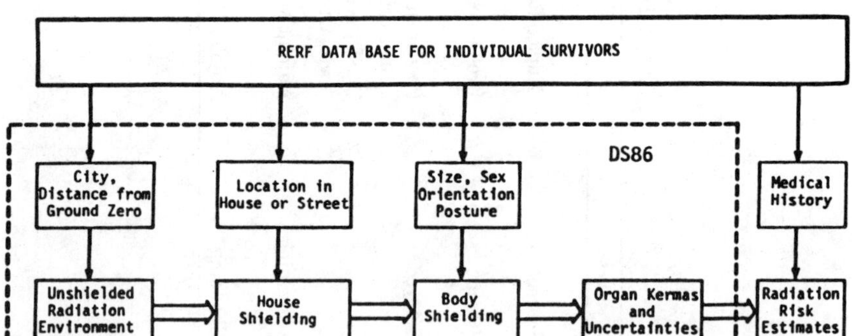

Fig. 1. The DS86 Interface and Operation on RERF Data for A-Bomb Survivors.

The kerma is calculated by

$$K = \int K(E)\phi(E)\mathrm{d}E$$

where $K(E)$ is the soft tissue kerma response function derived by Kerr [2]. For marrow, the response function is modified to account for the effect of the microstructure of bone on the dose in the marrow. The output provided by DS86 is given in Table 2. The neutron and gamma-ray fluences are summations over all neutron and gamma-ray components. This set of kermas and fluence should be of most use at RERF. *Notice, however, that the design of DS86 permits simple modifications in order to extract information in the form of other functions of the fluence that may be desired.*

The DS86 has been structured to accommodate propagation of uncertainty through the stages of the dosimetry and to account for correlation among kerma components. The model uses only fractional standard deviations (FSD), defined as the ratio of the standard deviation of the uncertainty distribution to the computed mean value, and correlation coefficients. It does not make assumptions concerning distribution of the uncertainty such as normal or log normal. Input to the model, at the present time, is largely based on uncertainties reported by the various investigators and judgements where necessary by the authors of DS86.

Sample runs of DS86 with RERF survivor data result in total FIA kerma FSDs of about 16% at Hiroshima and 13% at Nagasaki with the city difference attributed to the yield uncertainty. The organ kerma FSD with values in the 25% to 35% range is dominated by the house shielding location uncertainty. That is, the largest identified contributor to uncertainty in the organ kerma estimates was found to be the uncertainty in the shielding configuration as specified in the RERF computer files. Since a complete description of the shielding is available on paper, this uncertainty could be reduced with additional effort to code the survivor configuration into a more precise form usable by DS86. (See Chapter 7, of Reference 2.)

Risk analysis performed by RERF using the A-bomb dosimetry involves combining kerma estimates for individual survivors into cohorts. There are correlations in the dosimetry uncertainty among survivors grouped in these cohorts. These correlations depend on whether the survivors are in the same city or in different cities. Estimates of survivor-to-survivor correlations for the kerma factors have been made (Chapter 9 of Reference 2).

TABLE 2—*Output From DS86.*

Free Field	House Shielded	Body Shielded
Prompt Neutron Kerma	Shielded Prompt Neutron Kerma	Organ Prompt Neutron Kerma
Early Gamma-Ray Kerma*	Shielded Early Gamma-Ray Kerma	Organ Early Gamma-Ray Kerma
Late Gamma-Ray Kerma**	Shielded Late Gamma-Ray Kerma	Organ Late Gamma-Ray Kerma
Delayed Neutron Kerma	Shielded Delayed Neutron Kerma	Organ Delayed Neutron Kerma
Neutron Fluence	Shielded Prompt Neutron House Secondary Gamma-Ray Kerma	Organ Prompt Neutron House Secondary Gamma-Ray Kerma
Gamma-Ray Fluence	Shielded Delayed Neutron House Secondary Gamma-Ray Kerma	Organ Delayed Neutron House Secondary Gamma-Ray Kerma
	Shielded Neutron Fluence	Organ Prompt Neutron Body Secondary Gamma-Ray Kerma
	Shielded Gamma-Ray Fluence	Organ Delayed Neutron Body Secondary Gamma-Ray Kerma
		Organ Neutron Fluence
		Organ Gamma-Ray Fluence

*"Early Gamma-Ray" includes prompt gamma rays and prompt neutron air/ground secondary gamma rays

**"Late Gamma-Ray" includes fission product gamma rays and delayed neutron air/ground secondary gamma rays

Potentially, DS86 treats any survivor exposed within 2,500 m ground range of the bomb who was either in the open unshielded, in a Japanese residence or tenement house, or in the open shielded by a house. In the near term, models to calculate dosimetry for survivors who were shielded by terrain or who were in a factory will be added to the dosimetry system. However, most of the coded shielding data only exists for survivors in the so-called "proximally" exposed class, which includes survivors within 1,600 m ground range at Hiroshima or within 2,000 m ground range at Nagasaki. A tabulation of the proximally exposed survivors in RERF Life Span Study whose dose can be computed by DS86 is given in Table 3. Excluded from DS86 calculations are survivors in heavy shielded structures, such as concrete buildings, and other classes of survivors with unusual shielding configurations, such as survivors in streetcars or air raid shelters ATB.

The DS86 software design facilitates modification to the system to update or replace existing data bases or to expand its capabilities. Each of the data bases in DS86 are stand-alone modules with well-defined interfaces. One data base can be changed or completely replaced without disturbing the rest of the code or data. Additions to the systems can be "plugged in" at the appropriate point. Evidence for this facility was the ease in which a preliminary prototype dosimetry called DS84 installed at RERF was modified to become DS86, even though all of the data bases for DS86 were recalculated and expanded from those in DS84.

To illustrate the revised dosimetry, we discuss DS86 results for an example case from the RERF data base. This case was a male inside a house at 1,485 m ground range at Hiroshima. He was standing and facing sideways to the bomb direction.

The history file plan view for the case (left-hand side of Figure 2) shows the survivor on the bottom floor of a two-storied house. Please note the box enclosing the survivor location was added to emphasize that location and does not represent shielding. The arrows in the drawing point toward ground zero. The darkened portion of the circle pinpointing the location represents the face and indicates that the survivor was facing sideways to ground zero which was properly recorded in the computer data base. The DS86 shielding model for this location is a composite of several locations in the house cluster geometries used in producing the shielding data base. One of these locations and orientations relative to ground zero that is part of the composite is shown on the right-hand side of Figure 2. It is a reasonable configuration compared to the actual case. The location and

TABLE 3—*Proximal Exposed Treated by DS86.*

Shielding Configuration	Method	Number of Cases		Total
		Hiroshima	Nagasaki	
DS86				
Open—No Shielding	FIA	998	759	1,757
Japanese House	9 Parameter	7,385	2,380	9,765
Tenement House	9 Parameter	3,779	740	4,519
Open Shielded House	Globe	2,606	494	3,100
Totals, DS86		14,768	4,373	19,141
Update (1987)				
Terrain Shielded	Globe	0	315	315
Factory	Factory	25	664	689
Totals, Update		25	979	1,004
Not Treated (*e.g.*, Concrete Building, Air Raid)		525	1,084	1,609

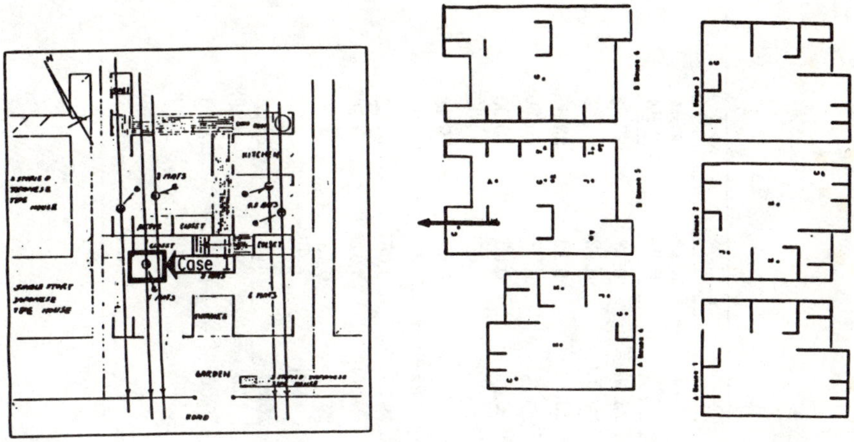

SURVIVOR HISTORY

DS86 MODEL (TYPICAL CASE FOR CATEGORY)

Fig. 2. Example Survivor Case Treated by DS86.

orientation combinations in the DS86 model for this survivor specification all place the survivor on the ground floor of a two-story house and relatively close to the wall facing ground zero.

The kermas computed at each stage and for the fifteen organs are presented in Table 4. The number in parentheses is the fractional standard deviation of the estimate. Note that the case example is male and therefore the female related organ dosimetry have zero values. The total kerma, total neutron and total gamma-ray kerma are presented. The house shielding causes a significant reduction in the radiation environment. The breakdown of the kerma values for the FIA, house shielded, marrow, and brain are shown in Table 5. Kermas from DS86 are compared to T65D for this case in Table 6. The small house shielding for gamma rays in T65D and the higher T65D neutron values for Hiroshima are apparent.

Finally, Table 7 contains a comparison of DS86 to T65D by presenting the ratio of DS86 kerma to T65D kerma at 1,400 m ground range at Hiroshima and Nagasaki for each dosimetry stage, and at 700 m and 2,500 m for the marrow. These values are based on average shielding by house and body at the appropriate ground range. The explanations for these ratios are well understood and are discussed in the dosimetry final report [2]. The points to emphasize here are that the differences between DS86 and T65D are significant and that these differences cannot be captured by a single factor—they depend on which stage (FIA, in-house, in-organ), which ground range, and which city. Thus, it is not appropriate

TABLE 4—*Kermas for Example Case.*

	Kerma (CGY) Total (FSD)	Kerma (CGY) Total N (FSD)	Kerma (CGY) Total G (FSD)		Kerma (CGY) Total (FSD)	Kerma (CGY) Total N (FSD)	Kerma (CGY) Total G (FSD)
FIA	52.66 (.16)	0.93 (.18)	51.73 (.17)	Liver	23.96 (.25)	0.13 (.32)	23.83 (.25)
Shielded	31.18 (.23)	0.37 (.27)	30.80 (.23)	Lungs	23.99 (.25)	0.12 (.32)	23.87 (.26)
Skeleton	22.17 (.25)	0.13 (.30)	22.04 (.25)	Ovaries	0.00 (.00)	0.00 (.00)	0.00 (.00)
Marrow	24.50 (.25)	0.11 (.30)	24.38 (.25)	Pancreas	19.53 (.25)	0.05 (.32)	19.48 (.25)
Bladder	23.61 (.26)	0.08 (.32)	23.53 (.26)	Stomach	18.90 (.26)	0.08 (.32)	18.82 (.26)
Brain	25.63 (.25)	0.15 (.32)	25.48 (.25)	Testes	24.71 (.26)	0.13 (.32)	24.58 (.26)
Breasts	0.00 (.00)	0.00 (.00)	0.00 (.00)	Thyroid	26.89 (.25)	0.17 (.32)	26.72 (.25)
Eyes	29.35 (.26)	0.23 (.32)	29.12 (.26)	Uterus	0.00 (.00)	0.00 (.00)	0.00 (.00)
Large Intestine	21.96 (.25)	0.05 (.32)	21.90 (.25)				

TABLE 5—*Kerma Components for Example Case.*

	Kerma Values (CGY)			
	FIA	House Shielded	Marrow	Brain
Neutron				
Prompt	0.917	0.368	0.1104	0.1508
Delayed	0.020	0.006	0.0013	0.0019
Total Neutron	0.927	0.374	0.112	0.153
Gamma Ray				
Early	23.4	14.9	11.8	12.5
Late	28.3	15.7	12.2	12.7
Prompt Neutron House Secondaries	—	0.18	0.14	0.142
Delayed Neutron House Secondaries	—	0.01	0.008	0.008
Prompt Neutron Body Secondaries	—	—	0.16	0.145
Delayed Neutron Body Secondaries	—	—	0.008	0.008
Total Gamma	51.7	30.8	24.4	25.5
Total Neutron and Gamma	52.6	31.2	24.5	25.6

TABLE 6—*Comparison of DS86 Dosimetry to T65D Dosimetry for Example Case.*

	Kerma (CGY)					
	Total		Total Gamma		Total Neutron	
	DS86	T65D	DS86	T65D	DS86	T65D
Free Field (FIA)	52.6	36	51.7	24	0.93	12
House Shielded	31.2	24.4	30.8	21	0.37	3.4
Marrow	24.5	14	24.4	13	0.11	0.95

TABLE 7—*DS86/T65D Kerma Ratio.*

	1400 M Ground Range			Marrow @ Ground Range	
	Free-Field	House Shielded	Marrow	700 M	2500 M
Hiroshima					
Neutron	0.08	0.11	0.13	0.23	0.10
Gamma Ray	2.1	1.1	1.5	0.96	5.2
Total	1.4	0.98	1.4	0.86	5.1
Nagasaki					
Neutron	0.38	0.42	0.53	0.64	0.56
Gamma Ray	0.82	0.46	0.63	0.86	0.57
Total	0.82	0.46	0.63	0.85	0.57

to make statements about changes in the radiation risk estimates based on simple scaling. We have to let RERF do their work with the new dosimetry, DS86, to find out.

References

1. Auxier, J.G., (1977), "Ichiban: Radiation Dosimetry for the Survivors of the Bombings of Hiroshima and Nagasaki," ERDA Critical Review Series, TID-27080, NTIS.
2. Roesch, W.C., Ed., (1987), "The Dosimetry of the Atomic Bombs at Hiroshima and Nagasaki," Report of a Binational Collaborative Study, Radiation Effects Research Foundation, Hiroshima, Japan.

The Use of DS86 for the Computation of Dose Estimates for Japanese A-bomb Survivors

Dale Preston*
Radiation Effects Research Foundation

Estimates of the risk of radiation effects in Japanese A-bomb survivors are crucial to our understanding of the health effects of radiation and to attempts to set standards for radiation protection. In order to produce reliable estimates of the risks associated with radiation exposure among A-bomb survivors it is important that we have good estimates of the doses received by individual survivors. In the late 1970's and early 1980's a number of physicists raised questions [1,2] about the tentative 1965 dosimetry system (T65D) which had been used by the Radiation Effects Research Foundation (RERF) and its predecessor, the Atomic Bomb Casualty Commission (ABCC) to compute dose estimates for individual survivors. As a result of these questions a joint US-Japan committee was established to review all aspects of the dosimetry for Hiroshima and Nagasaki atomic bombs. In April 1986, after over four years of work, the committee approved a new dosimetry system, christened DS86, for use by RERF.

Following a short history of dose estimation at RERF, I will briefly describe the new dosimetry system and discuss how this system has been used by RERF to compute organ dose and exposure kerma estimates for individual survivors. I will also very briefly summarize some early results on how these new dose estimates will affect the estimated risks for leukemia and nonleukemia non-cancer mortality.

*Currently with the Radiation Epidemiology Branch, National Cancer Institute

These results are preliminary. RERF is in the process of preparing a series of reports which will discuss in detail the various topics I will mention here.

Many people at RERF have been involved in the work on procedures for the use of DS86. I would particularly like to acknowledge Mr. S. Fujita of the Statistics Department and Mr. H. Ebisui of the Computer Center for their contributions. The results for leukemia and nonleukemia cancer mortality discussed below are based on analyses carried out by Professor D.A. Pierce and myself.

Terminology

In this presentation, the following terms will be used to describe various aspects of the dosimetry. *Free-in-air (FIA)* kerma refers to the tissue kerma in air at a survivor location unadjusted for shielding by external structures or terrain. Tissue kerma in air adjusted for the effects of shielding will be called *shielded kerma* or, more simply, *kerma*. The mean absorbed dose for a specific organ will be called *organ dose* or *dose*.

Dosimetry Systems Used by ABCC and RERF

In the late 1950's a dosimetry system for persons exposed to A-bomb radiation in Hiroshima and Nagasaki was made available for use at ABCC. This dosimetry is generally referred to as the tentative 1957 dosimetry or T57D. T57D uses the York "air dose" curves to describe the gamma and neutron FIA kerma versus distance relationships for the two cities. These curves were based upon data obtained from weapons tests and assumptions about the yield of Hiroshima and Nagasaki bombs. The only information in the T57D system about shielding by structures or terrain were crude attenuation curves for gamma rays and neutrons as a function of the distance from the person to a point on the structure which was in a direct line to the epicenter. The T57D system contained no information about organ dose. The most complete description of T57D is given by Ritchie and Hurst [3].

The T57D system was never regarded as a definitive dosimetry. The primary use of the T57D system by ABCC investigators was to identify people who were believed to have significant exposures (roughly this meant T57D FIA kerma estimates in excess of about .15 Gy). No systematic program for the computation of individual T57D estimates for members of the Life Span Study (LSS) cohort was ever undertaken.

Because of the recognized weaknesses of the T57D system, efforts were begun in both Japan and the US to develop a better dosimetry for the A-bomb survivors. These efforts focused on refinement of the York curves and the development of better procedures for estimation of shielded kerma for survivors who were known to have been inside houses or other light structures at the time of the bomb (ATB). This work, which has been described by Auxier [4], lead to the development of the tentative 1965 dosimetry (T65D) system. In the late 1960's the T65D system was adopted as the official dosimetry system for ABCC studies and a program was undertaken to compute kerma estimates for all survivors in the various ABCC mortality samples. T65D estimates have been computed for 115,025 survivors. With minor revisions, T65D estimates have been used in all ABCC/ RERF analyses since about 1970. Milton and Shohoji [5] discuss the use of the T65D system by ABCC.

In order to understand how RERF has used the DS86 system, it is necessary to review briefly the data and procedures used in the computation of T65D estimates. T65D FIA kerma estimates are computed using city-specific parametric functions which describe the FIA kerma versus distance relationships. For both gamma rays and neutrons these functions have the form

$$k = G_0 e^{-r/L} / r^2$$

where r is the distance from the survivor to the epicenter and G_0 and L are parameters specified by the T65D system. T65D FIA kerma estimates have been computed for all survivors whose distance from the hypocenter was known.

T65D gamma-ray and neutron transmission factors are assigned, based upon information about shielding category and distance for individual survivors. T65D kerma estimates are computed as the product of the FIA kerma components and the appropriate transmission factors. The methods used to estimate T65D transmission factors for survivors with detailed shielding histories are: i) the nine parameter method in which the factors are computed as linear functions of nine parameters which describe the location of the survivor inside a house or similar structure; ii) the globe method in which the amount of shielding was determined by direct observation of shadows cast by a light placed in a model of the survivor's environment ATB; and iii) ad hoc assignment of transmission factors based on a review of individual histories or groups of similar histories.

The globe method was used for survivors who were outside but completely or partially shielded by houses or terrain and for some survivors (in Nagasaki) who were in factory buildings ATB. Two important examples of ad hoc assignment of transmission factors are: i) the choice of factors for most Nagasaki factory workers and ii) the use of average house transmission factors for survivors without detailed shielding histories who are known to have been in houses ATB.

Detailed shielding information is generally available only for survivors who were within 2,000 m of the hypocenter in Nagasaki or 1,600 m in Hiroshima (proximal survivors). In the T65D system, FIA kerma was used as the kerma estimate for all survivors who were further from the hypocenter ATB (distal survivors).

The original T65D system did not include information about organ dose. Although virtually all ABCC and RERF analyses based on T65D estimates have used kerma as the basic dosimetric quantity, several recent analyses have made use of organ doses computed using the gamma ray, neutron, and neutron-capture gamma-ray transmission factors for selected organs given by Kerr [6] in 1979.

Direct DS86 Estimates

DS86 is considerably more complex than T65D. In particular, estimates of FIA kerma, kerma, and organ dose are all derived from detailed descriptions of the energy and angular fluence distributions for gamma rays and neutrons at the survivor location. The system provides much more detailed dosimetric information than can be obtained from T65D.

Unlike the T65D system, DS86 does not provide FIA kerma estimates as simple functions of the distance from the epicenter. Nor does the system make explicit use of transmission factors in the direct computation of kerma or organ dose estimates.

The DS86 system can be used directly to compute kerma and organ dose estimates for survivors with detailed shielding histories in selected shielding categories who were within 2,500 m of the hypocenter. The shielding categories considered in the current version of DS86 are: i) survivors who were inside houses or other light wooden structures; ii) survivors who were outside but completely or partially shielded by houses; and iii) survivors who were in the open and unshielded.

For survivors in the above categories, the DS86 program first computes the distance from the hypocenter. If this distance is less than 2,500 m the appropriate fluence distributions are extracted from a fluence data base. These distributions are then coupled with soft tissue conversion factors to produce FIA kerma component estimates. Next, the shielding data are used to determine the DS86 shielding case which best describes the external shielding conditions for the survivor. Based on the shielding case, data on appropriate transfer functions are extracted from the data base and used to adjust the estimated fluence distributions. Kerma components are then computed from these adjusted distributions. For a person who was in the open and unshielded, the FIA kerma and kerma components are identical. Because of questions about the accuracy of the shielding information DS86 estimates have not been computed for survivors who, according to the shielding history, were in the open and unshielded but for whom there was no evidence of flash burns.

For each organ of interest the appropriate transfer function data are extracted from the DS86 data base and used to estimate the fluence at the organ. The organ dose response function is then used to compute the estimates of the average absorbed organ dose components.

The direct computation of kerma and a full set of organ dose estimates (about 13) requires over two minutes per person on the moderately fast mainframe computer currently in use at RERF. About two and a half months of computer time was required to compute direct DS86 estimates for 23,427 survivors. This group accounts for 82% of the survivors for whom detailed shielding data are available.

Indirect DS86 Estimates

As noted above, T65D estimates have been computed for 115,025 of the survivors in various RERF samples. However, detailed shielding data are available for only about 25% of these survivors. In order to provide DS86 estimates for survivors without detailed histories it has been necessary to develop indirect estimation procedures.

For those proximal survivors whose T65D kerma estimate was computed by application of average T65D house transmission factors, DS86 kerma estimates were computed in a similar manner. That is, average house transmission factors computed (post hoc) from persons with direct DS86 estimates were applied to DS86 FIA kerma component estimates. The FIA kerma estimates for these survivors and for the more distal survivors discussed below were computed using city-specific parametric models of the DS86 FIA kerma versus distance relationships developed at RERF. The form of these models is similar to the T65D FIA kerma models.

Because of the increased shielding provided by houses and other light structures it was felt to be inappropriate to continue the T65D practice of using the FIA kerma estimate as the shielded kerma estimate for distal survivors. Since little is known about other shielding situations it was decided to limit the computation of DS86 estimates for distal survivors to those whose total DS86 FIA kerma estimate is small (<5mGy) or who were, according to ABCC records, in houses ATB. Persons in the former group were assigned DS86 kerma estimates of 0 Gy, while average DS86 house transmission factors were used to compute kerma estimates for survivors in the latter group. Organ dose estimates for all survivors with indirect DS86 estimates were computed using city-specific DS86 average organ transmission factors computed from survivors for whom direct estimates could be made.

Table 1 contains a summary of the current status of the DS86 computations. The first column in this table contains information about the distribution of estimate types for members of the LSS cohort which forms the basis of most RERF studies. The second column contains similar information for members of other samples for whom efforts have been made to compute T65D estimates.

TABLE 1—*Summary of the current status of the computation of DS86 estimates.*

Estimation Method	Cohort		
	Life Span Study	Others	Total
Direct DS86	18,524	4,896	23,420
Indirect DS86	57,767	13,580	71,417
DS86 Total	76,091	18,746	94,837
Not-in-City	16,580	30	26,610
No DS86 Estimate	17,650	2,538	20,168
Total	120,321	21,314	141,635

Comparison of T65D and DS86 Estimates

In general terms, as illustrated in Figure 1, DS86 gamma FIA kerma estimates in Hiroshima are about 40% higher than the corresponding T65D estimates. This difference increases with distance.

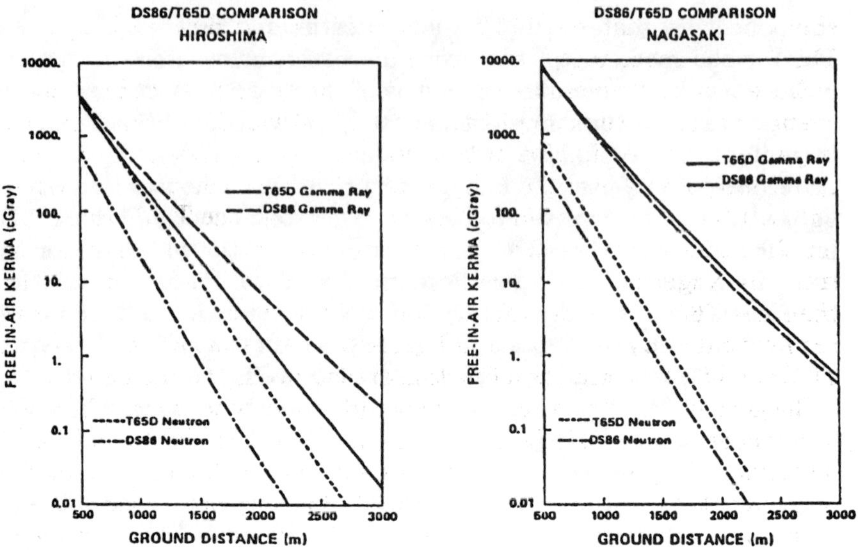

Fig. 1. DS86 and T65D FIA kerma versus distance curves for Hiroshima and Nagasaki.

The Hiroshima DS86 FIA neutron kerma estimates are about 10% of the old values. In Nagasaki, there is little change in the FIA gamma kerma estimates while the neutron estimates are reduced by about 30%. The inter-city difference in the fraction of the total dose derived from neutrons is reduced with the new dosimetry.

Table 2 contains observed values of the average gamma and neutron house transmission factors for survivors with direct estimates who were more than 1,000 m from the hypocenter. These values indicate that the DS86 average gamma house transmission factor is

TABLE 2—*Comparison of DS86 and T65D transmission factors for Japanese-style house and other light structures. The DS86 values were computed as the average of the appropriate ratios for all survivors beyond 1,000 m from the hypocenter with direct DS86 estimates.*

City	Dose Component	Dosimetry	
		DS86	T65D
Hiroshima	gamma	0.46	0.90
	neutron	0.36	0.32
Ratio of gamma factors(DS86/ T65D) = 0.51			
Nagasaki	gamma	0.48	0.81
	neutron	0.41	0.35
Ratio of gamma factors(DS86/ T65D) = 0.59			

about 50% of that for T65D in Hiroshima and 60% in Nagasaki. There is little change in the factors for neutrons.

Since many survivors were inside houses or similar structures these changes in the estimates of the shielding provided by houses have a large impact on estimates of shielded kerma. In particular, the average DS86 gamma kerma estimates for Hiroshima survivors who were within 1,200 m of the bomb is slightly less than the T65D estimates for the same survivors, while for more distal survivors the reverse is true. In Nagasaki, the DS86 gamma kerma estimates are smaller than the T65D estimates for virtually all survivors. These findings are illustrated by the curves in Figure 2 which contrast the average T65D and DS86 kerma as a function of distance in the two cities.

Because of differences in the geography and time of day there are differences in the shielding conditions in Hiroshima and Nagasaki. In particular, the proportion of survivors who were in houses is lower for Nagasaki than Hiroshima. Because DS86 does not currently deal with some of the important shielding situations in Nagasaki, e.g., factories and terrain shielding, the proportion of people for whom DS86 estimates cannot yet be computed is higher in Nagasaki (22%)

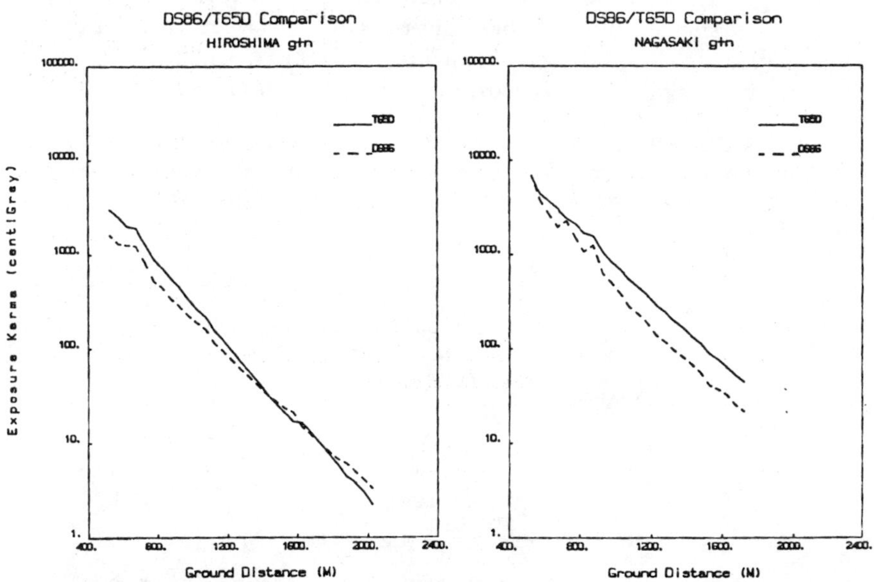

Fig. 2. Average kerma estimates versus distance for all survivors in the Life Span Study Cohort for whom DS86 estimates have been computed. The curves shown were produced by averaging the total kerma for all survivors in 50 m intervals of ground range.

than in Hiroshima (17%). If attention is limited to survivors whose T65D kerma estimates are between 0.5 and 2 Gy this inter-city contrast is even more striking. DS86 estimates cannot be computed for about 5% of the Hiroshima survivors in this group while the corresponding figure for Nagasaki is about 40%.

Comparison of the observed DS86 average gamma transmission factors for various tissues with the earlier estimates given by Kerr indicates that for active bone marrow and deep seated organs such as the stomach or large intestines the DS86 factors are about 80% greater than they were previously thought. The differences for more superficial tissues, such as the female breast of the thyroid, are less marked. Table 3 contrasts observed average DS86 factors with Kerr's earlier values for several representative tissues.

Comparison of T65D and DS86 Estimates of the Risk of Cancer Mortality

This section contains a brief summary of some results on cancer mortality. These estimates are based on cancer mortality in the RERF Life Span Study cohort during the period 1950 through 1985. Separate results are given for all cancers except leukemia as a group and for leukemia. This is done because both the level and temporal pattern of the radiogenic risks differ greatly between these. Reports which discuss the findings outlined here in more detail and the results for analyses of specific sites are currently being prepared.

TABLE 3—*Comparison of DS86 and T65D transmission factors for selected tissues. The DS86 values were computed as the average of the appropriate ratios for all survivors beyond 1,000 m from the hypocenter with direct DS86 estimates.*

Tissue	Dose Component	Dosimetry	
		DS86	T65D
Bone Marrow	gamma	0.81	0.56
	neutron	0.37	0.24
	n-gamma	0.42	0.07
Ratio of gamma factors(DS86/ T65D) = 1.45			
Large Intestine	gamma	0.74	0.40
	neutron	0.19	0.14
	n-gamma	0.41	0.08
Ratio of gamma factors(DS86/ T65D) = 1.85			
Female Breast	gamma	0.87	0.80
	neutron	0.67	0.55
	n-gamma	0.28	0.05
Ratio of gamma factors(DS86/ T65D) = 1.09			

Preliminary results for both leukemia and all cancers except leukemia indicate that the primary effect of the change in dosimetry is to change the slope of the dose response curve. Variations in risk by sex, age-at-exposure, and time-since-exposure have been found to be virtually identical for both dosimetries.

The effect of the change in dosimetry on the level of risk depends upon whether the estimates are based upon kerma or organ dose and upon the assumed value of the RBE for neutrons. For both leukemia and all cancers except leukemia, the risks per Gy based upon total DS86 kerma estimates are 60-70% greater than the corresponding T65D risks. In terms of total organ dose (RBE = 1), using dose to the marrow for leukemia and dose to the small intestine as a representative dose for other cancers the DS86 and T65D risk estimates are about equal. However, if an RBE of 10 is assumed the DS86 leukemia risk estimate is 80% greater than the T65D estimate and the DS86 nonleukemia risk is about 30% greater than the corresponding T65D estimate. The differences between the DS86 and T65D risk estimates increase if one assumes larger values for the RBE.

It should be noted that because of the significant reduction in the proportion of the total dose due to neutrons with the DS86 dosimetry, particularly for Hiroshima, the new risk estimates are much less sensitive to assumptions about the RBE than were the T65D risk estimates. Our work with the DS86 dosimetry suggests that it will be impossible to use the LSS data as a basis for the estimation of the RBE of neutrons.

References

1. Loewe, W.E., Mendelsohn, E.: Neutron and gamma doses at Hiroshima and Nagasaki. *Health Phys.* 41:663, 1982.
2. Bond, V.P., Thiessen J.W. (eds.): *Reevaluation of Dosimetric Factors: Hiroshima and Nagasaki.* DOE Symposium Series 55, CONF-820928, 1982.
3. Ritchie, R.H., Hurst, G.S.: Penetration of Weapons Radiation: Application to the Hiroshima and Nagasaki Studies. Atomic Bomb Casualty Commission Technical Report 26-59, 1959.
4. Auxier, J.A.: *Ichiban—Radiation Dosimetry for Survivors of the Bombings of Hiroshima and Nagasaki.* ERDA Critical Review Series, TID-27080, NTIS 1977.
5. Milton, R.C., Shohoji, T.: Tentative 1965 radiation dose estimation for atomic bomb survivors, Hiroshima and Nagasaki. Atomic Bomb Casualty Commission Technical Report 1-68, 1968.
6. Kerr, G.D.: Organ dose estimates for Japanese atomic bomb survivors. *Health Phys.* 37:487-508, 1979.

Discussion

L. WAGNER: I'd like to ask one of the speakers about the effects of posturing on the calculations. Approximately how much change in absorbed dose would there be depending on the different postures?

D. KAUL: The biggest effect we saw was, of course, prone vs. standing where you could get 50 percent differences, where you get a lower dose for prone, particularly for some organs which are asymmetrically distributed in the body. For the kneeling or sitting position you could get changes of 20 to 30 percent in a range of perhaps 10 to 30 percent differences from a standing phantom. These differences, particularly for the sitting vs. standing phantom, really became important only if we knew what the orientation of the survivor was because, particularly with some non-uniformly distributed or asymmetrically distributed organs, the change, just in orientation relative to the hypocenter, whether the man was facing it, or facing away or facing to the side, were of the same order as the differences between sitting and standing.

L. WAGNER: I think you answered another question I had which was that the second speaker showed us slides about the thyroid and breast doses and I was going to ask what the orientation effect relative to the hypocenter would be because those organs are external if you're facing it, but if you're facing away, then they're at the depth dose and that should change the dosimetry a little bit.

D. PRESTON: I just want to comment that those kinds of effects are fully accounted for in the dosimetry system for the individual survivors as that data is known. I'll just say that the values I showed are averages for all people so they represent sort of a random orientation.

E. RADFORD: Dr. Preston, would you comment on the fact that you obtained apparently an excess relative risk of 0.7 or 70 percent per survivor. If I read your figure correctly, that this is for all cancers except leukemia, I calculate a very similar excess relative risk. This implies a doubling dose of approximately 130 rads from the Japanese A-bomb data. Do you want to comment on that?

D. PRESTON: Based upon the calculations we've done, using an excess risk model for leukemia, allowing the risk to decrease in time, so it's not the standard constant excess risk model, we do get a slope in dose of about 0.7 like you said.

E. RADFORD: Translated on the assumption of a 10 percent death rate overall of the Japanese from cancer as a whole, other than leukemia, that works out to well over 1,000 excess cases per million persons per rad.

W. LOWDER: I'd like to ask Bill Woolson to comment a little bit on the significance of the fact that we have the fluence in some detail as a function of energy within the organs as one of the outputs of the DS86. Right from the beginning I think all of us were looking ahead to the fact that we would want to apply this information to more detailed microscopic-scaled calculations and perhaps a comment on this plus an indication of whether you have experimented at all with the applying of an additional secondary particle code to the existing system.

W. WOOLSON: Well, as I mentioned in the talk, the fluences are provided at the spatial dependent points within the various organs and that information is available in the dosimetry system. As far as us carrying that calculation further with some kind of straggling calculation we have not done any of that. The marrow-dose response function that is in the system to calculate the kerma, does include the effect of micro bone structure as George Kerr talked about earlier. So these types of calculations certainly can be carried out but we haven't done anything on that yet.

R. WITHERS: As I remember the curve from this morning, the dose was halving about every 100 m distance from the center of the explosion. How accurately are the distances of the survivors known from that center and how were they validated?

D. PRESTON: The map coordinates in the computer that are stored in the data base for the survivors were taken off US Army maps and they were recorded generally to the nearest 10 m. That is what we have in the data base. For some survivors, especially those that are more distal, there is indication that they were actually recorded to the nearest 100 much more often than you would like. So, it is likely that this is a major source of error in the dosimetry, not in the dosimetry system, but in the dose estimates that are produced. And I think it is very important to remember that even though we do have sources of variability in the dosimetry system, there are a lot of other sources of variability in the RERF data as it was recorded over the years and I think Mr. Jablon, in the early 1970's, indicated that an estimate of the error of about 30 percent would not be unlikely. This does not include the kinds of errors that Dr. Woolson and Dr. Kaul talked about earlier today.

R. WITHERS: You say an error of 30 percent. You mean in dose or in distance?

D. PRESTON: In the dose estimates.

E. RADFORD: If I could just comment on that. Mr. Jablon also pointed out that the effect of that uncertainty would be to lessen the apparent slope of radiation effect on dose.

N. ROHRIG: I've been intrigued by the new mental retardation data and I think it would be very useful in the DS86 data to split the conventional 10 to 50 rad group into a group of 10 to 30 rads and another group of 30 to 50 rads. Is there any plan to do this?

D. PRESTON: In fact, in the analysis I reported, I showed some tables that used the conventional categories because they were familiar but actually we used a slightly finer grouping and some of the categories were similar to what you described. Those categories are also being used in the analysis for Report 11 and I think will become fairly widely used. There is also consideration in looking at dose response in the range of about 10 to 200 rads. There is a need for more detail if we hope to make any more inferences; it's going to be difficult anyway.

R. CHRISTY: I'd like to ask Dr. Preston about the category that you described as in the open, unshielded but I believe you said showed no flashburns. I did not quite understand this.

D. PRESTON: I was talking there about individuals who were within 2,000 m and, in Hiroshima, within 1,600 m for whom shielding histories are available and there are also indications of acute symptoms. One of the questions was "Did you have any burns?" For those people who said there was no indication of any burns, we did not, even though in T65D they were given a transition factor of 1 essentially. So they were assigned a free in air kerma and for the present time, in the DS86 calculation, we have not included those people.

W. WOOLSON: Dale, I have a follow-up question to Dr. Christy. Presumably, those people that didn't have any flashburns you might subdivide into two other classes: those that might be unreliable and those that may have in fact had additional shielding that wasn't captured in the T65 system. I was wondering whether you think it would be possible to examine the detailed histories for those survivors since they are in the proximal class and provide some kind of shielding assessment for them to put them back into the cohort.

D. PRESTON: I think something like that should be done. There is a need to look at more detail on some of these categories. And there are efforts under way for certain groups of survivors to do that and also, for example in the 9-parameter data, what kind of uncertainties are there, so as to review the quality of that data both in the coding aspects and the transcribing that went on.

D. KAUL: For the person who asked the question about the effects of orientation and posturing and what have you. There is a great deal of information in the final Report where the data is tabulated and compared for various ranges and for all sorts of different postures and orientations so anybody who is interested in those effects can get it out of that Report. Another point is the issue of the body n-gammas. It's interesting to know that these obviously will be affected if it turns out that cobalt activation and europium activation data are correct. And if, for some reason or other, something happens to the DS86 neutrons, so that that activation agrees, that is, the fluence goes up by a factor of 5 at 1,200 m, then the body n-gammas will go up by a factor of 5 at 1,200 m. The number that's there, which is a number on the order of 0.4 transmission from conversion from neutron rads to gamma rads, is a good number. The reason that the original Troice Jones data of 0.07 are so different is that Troice used a very fast spectrum to do his analyses and the number of neutrons per rad in that spectrum was relatively small. With the Hiroshima environment, the number of neutrons per rad is very high, and hence the conversion to gamma rays is considerably more efficient.

E. WEBSTER: Dr. Preston in a previous answer suggested that the constant relative risk model was not holding up as a function of advancing age, and I wonder if you can enlarge on that. Maybe I misunderstood what you said.

D. PRESTON: I think I was talking about leukemia when I was talking about non-constant risk. And I was saying that we used a model in which we modeled the excess risk directly and allowed that to decrease with time. As far as relative risk, if you look at cancers other than leukemia and most individual non-leukemia cancers, there is very little evidence against the constant risk model.

Lauriston S. Taylor Lectures in Radiation Protection and Measurements

Lecture No. 11

How to be Quantitative about Radiation Risk Estimates

by Seymour Jablon

Presented April 8, 1987
Issued March 15, 1988

National Council on Radiation Protection and Measurements
7910 WOODMONT AVENUE/BETHESDA, MD. 20814

Lauriston Sale Taylor

THE LAURISTON S. TAYLOR LECTURES IN RADIATION PROTECTION AND MEASUREMENT

These lectures are intended to honor Lauriston S. Taylor who was the first President of the National Council on Radiation Protection and Measurements (NCRP) from its inception in 1964 until he retired in 1977. Prior to 1964, he was chairman of all the NCRP's predecessor national committees beginning in 1929. A detailed account of his career and achievements is contained in an appreciation which appears in Lecture No. 1 of this lecture series.

The lectures are given once annually by distinguished lecturers selected by the Program Committee of the NCRP. They treat a topic related to radiation protection or measurement. The lectures are sponsored by the NCRP, are held on the occasion of its Annual Meeting and are published subsequently under the auspices of the Council.

Initiation of the lecture series was made possible by a generous grant from the James Picker Foundation. For this the Council expresses its deep appreciation.

The Eleventh Lauriston Taylor lecturer, Mr. Seymour Jablon of the National Research Council of the National Academy of Sciences, will be introduced by his distinguished friend and colleague, Dr. Gilbert Beebe of the National Cancer Institute, National Institutes of Health, Bethesda, Maryland, a well known epidemiologist and a good friend to the NCRP. Dr. Beebe . . . .

Warren K. Sinclair
President

Seymour Jablon

SEYMOUR JABLON

Eleventh Lauriston S. Taylor Lecturer in Radiation Protection and Measurements

It is, of course, a privilege to introduce a Lauriston S. Taylor lecturer. But for me it is also a great personal pleasure because the lecturer today is Seymour Jablon, a friend and colleague with whom I worked at the National Research Council for almost 30 years, 1948–1977, minus the 12 years during which one or the other of us was in Japan on duty at the Atomic Bomb Casualty Commission. But even then we kept in close touch by correspondence.

Seymour Jablon was born in New York City in 1918. He attended the public schools, including the City College of New York from which he was graduated in 1939, majoring in mathematics. He obtained a master's degree in mathematics from Columbia University the following year, transferred his affections to mathematical statistics in 1940–1941, and during World War II served as a communications officer in the Army Air Forces. He was stationed mainly in India, helping the Air Transport Command when it was ferrying supplies over the hump to China. After the war he returned to Columbia as a Ph.D. candidate in mathematical statistics and supported his growing family by teaching statistics at Rutgers. In 1948, with only a dissertation to finish, he joined the professional staff of the Division of Medical Sciences, National Research Council, as a statistician in the Medical Follow-up Agency which had been established at the request of the Armed Forces and the Veterans Administration to conduct medical follow-up studies on WW II veterans. He distinguished himself in this non-radiation work through his knowledge of mathematical statistics and his skill in working with a wide variety of medical collaborators interested in tuberculosis, head wounds, hepatitis, epidemic hemorrhagic fever, hand surgery, schizophrenia, schistosomiasis, twin research, sarcoidosis, an influenza vaccine, and other topics.

In November, 1946, the National Academy of Sciences had received a commission from President Harry S. Truman to organize medical follow-up studies on the A-bomb survivors in Japan. Responsibility for the program fell to the Division of Medical Sciences which established the Atomic Bomb Casualty Commission (ABCC) in Japan in 1947. Although the Medical Follow-up Agency and the ABCC programs were quite separate, staff discussions exposed Seymour to the early evidence of excess leukemia among the A-bomb survivors and to discussions of the ABCC program, particularly its lack of an overall research strategy.

Stimulated by difficulties in recruiting Japanese scientists and by the paucity of evidence of early ill effects other than leukemia and cataracts, an AEC Advisory Committee for Biology and Medicine in 1951 recommended that ABCC be phased out by 1953. This was not done but, despite some new findings, dissatisfaction remained and by 1955 the situation was so acute that Dr. R. Keith Cannan, Chairman of the Division of Medical Sciences, asked Dr. Thomas Francis, the epidemiologist famous for his work on the influenza virus and his management of the Salk polio vaccine trials, to make a site visit to Hiroshima together with two statisticians, Seymour Jablon and Felix Moore who was then at the National Heart Institute.

The product of their site visit, known as the Francis Report,

Gilbert Beebe

was the turning-point in the ABCC program, providing an overall research strategy that has stood the test of time and changing trends in research. In 1957 Seymour and Felix Moore were back in Japan, arranging for the implementation of the Francis report. Also in 1957 Seymour persuaded me, as head of the Medical Follow-up Agency, to offer our services to the Division of Medical Sciences as one of the several stateside anchor groups Dr. Cannan planned to organize in support of the ABCC professional departments. This involved our taking responsibility for statistical aspects of planning, recruiting, and program management. Beginning in 1960 Seymour served two 3-year tours in Japan, first as head of the Department of Statistics, later as head of a consolidated Department of Epidemiology and Statistics. He proved to be an able, vigorous leader and his efforts are responsible for much of the excellence of the ABCC-RERF program. He took an active role in program planning, showed a critical interest in dosimetry, completed the accumulation of the Life Span Study sample, determined patterns of analysis and statistical methodology, and introduced the computer. Many of his papers are primary references on the A-bomb survivor experience. In 1975 he returned to ABCC as a member of the Crow Committee that reviewed the scientific program of ABCC for the NAS and made recommendations for its future conduct under the Radiation Effects Research Foundation (RERF) established on 1 April, 1975, pursuant to a formal agreement between Japan and the U.S.

Beginning in 1976 Seymour gradually assumed more and more of the responsibilities of a project officer for the Department of Energy (DOE) contract supporting RERF work and became Associate Director for International Affairs, Assembly of Life Sciences. These responsibilities have taken him to Japan 3 to 4 times a year and involved him in continuing negotiations not only with the Board of Directors of

RERF but also with its parent organization, the Japanese Ministry of Health and Welfare.

Recruiting top-flight U.S. personnel for the work in Japan has been a major concern. Recently the plunging value of the dollar against the yen has been very stressful for him as his budget for the U.S. support of RERF is prepared in dollars and there has been no protection against its changing value. Despite his heavy administrative duties his scientific contributions to the radiation literature have continued full force.

If the data coming out of the studies in Japan have great significance for radiation protection, as I believe they do and will increasingly, much of the credit is due to Seymour's scientific leadership in the ABCC period and to his stewardship during the past decade. In addition he has served you as a member of the NCRP Council 1974 to 1986, and as a mainstay on Committee 71. Throughout his career Seymour has been much sought after as a member of various advisory groups at NIH, DOE, and elsewhere. He also contributed heavily to the NAS BEIR I report and to the recent Report of the NIH Working Group To Develop the Radioepidemiological Tables.

Seymour has been the recipient of a number of honors, including a major award in 1975 from the National Academy of Sciences for distinguished service, and it is rumored that he will shortly receive a medal from the Emperor of Japan.

Seymour Jablon has done a great deal for our field and I am happy to introduce him as the Lauriston S. Taylor lecturer for 1987.

Gilbert Beebe
National Cancer Institute, NIH
Bethesda, Maryland

Seymour Jablon

How to be Quantitative About Radiation Risk Estimates*

by SEYMOUR JABLON

It is, of course, a great honor for me to have been chosen this year to help in paying tribute to Lauriston Taylor. His contributions to this organization and to radiation protection are so well known that description by me would be superfluous, and incomplete. I will content myself, therefore, with saying thank you, Laurie, for all that you have done.

Those of you who have surveyed the list of the previous Taylor Lectures will have recognized that my title was inspired by Bill Pochin's title for the Second Taylor Lecture, given nine years ago. For those who don't remember, his title was, "Why Be Quantitative About Radiation Risk Estimates?" What Bill said, in one sentence, was, essentially, that one can't reasonably propose limits of exposure unless one has at least a

* Presented April 8, 1987 in the auditorium of the National Academy of Sciences Building, Washington, DC.

rough notion of what risks may be implied by those limits. In a wholly rational world that would be true. In the real world, however, it seems to me that we have adopted limits to exposure that resulted from an inter-play between the things that we wished to accomplish by the use of radiation and the technical feasibility of accomplishing them under certain rules for the limitation of exposures. As a second step, the limits have been evaluated using available estimates of risk, however uncertain they might be. Most people have concluded that the risks while adhering to the occupational limits are very small, and should be considered acceptable. Whatever one's opinion concerning the rationality, or the reverse, of this mode of procedure, I think many of us would agree that the situation is far better than is true of the methods used in the United States to control, for example, the use as food additives of substances that may possibly be carcinogens for man.

John Lawrence gives an interesting account of how it was first determined that neutrons posed great risks:[1]

> Very little was known of the biological effects of the neutron rays produced by the cyclotron, and this seemed an important place to start work. For the neutron ray exposures in Berkeley we made a small metal cylinder to house a rat so that it could be placed close to the cyclotron. After placing the rat in position, we asked the crew to start the cyclotron and then turn it off again after the first two minutes . . . After the two minutes had passed, we . . . opened the cylinder, and found the rat was dead. Everyone crowded around to look at the rat, and a healthy respect for nuclear radiations was born . . . We discovered later that the rat's death had resulted from asphyxiation rather than radiation. But since our failure to aerate the rat chamber adequately had brought about such a salutary effect on the crew, the post-mortem report was not widely circulated.

In reading through the titles of previous Taylor Lectures in search of inspiration for this one, I was tempted to try a variation of Jim Crow's title for the Fifth Lecture, seven years ago. It was "How Well Can we Assess Genetic Risk? Not

Very." I think that we may be in a slightly better situation with respect to somatic risk than genetic, but Jim's title would be as appropriate for somatic risk as it was for genetic—as I will try to make clear.

I'm not going to have anything to say about the risks of genetic effects, for the simple reason that I know little more about them than what is contained in such documents as the BEIR and UNSCEAR reports. I'm going to limit myself mainly to a topic that is more than large enough for me—stochastic effects in man, that is to say, those effects whose probability of occurrence varies with the radiation dose, while the magnitude of the effect, if it does occur, does not depend upon the dose.

The principal stochastic, somatic, effect of ionizing radiation is, of course, the induction of certain cancers. Although it may be true that any form of cancer can be induced by radiation, certain forms, notably chronic lymphocytic leukemia, if inducible, are so rarely produced that it has never been possible to find excess numbers occurring in any irradiated population. Other somatic effects, such as those that occur in the lens of the eye or the developing fetal brain are not at all well understood and I will say only that it is most unlikely that mental retardation following fetal radiation is a stochastic effect; the magnitude of the deficit that is induced surely increases with the magnitude of the dose.

We will, then, be talking mainly about the induction by radiation of leukemia and some other cancers.

As I have said, Sir Edward Pochin, nine years ago, justified the need for quantification in radiation risk estimation on the basis that, in order for us to be able to be quantitative in a sensible way in setting limits to exposure, we need to have some idea about the magnitude of the risks that may be possible even if the dose limitations are adhered to scrupulously.

I believe, however, that this objective, important though it be, is only part of the reason that we want to be quantitative about risk estimates. There is another, deeper reason—to learn something about the process of radiation carcinogenesis and, perhaps, doing that, to learn something about the carcinogenic process itself. The time interval between a radiation exposure and the first recognition of a radiation induced cancer; the shape of the dose response curve; the way in which cancer risk varies with age, with sex and with time after exposure, the way radiation risks vary in the presence of other carcinogens (for example, cigarette smoking)—all these tell us something about what is going on biologically. At the least, any theory of radiation carcinogenesis must be consistent with the epidemiological data. At best, the epidemiological observations will suggest the biologic processes that must underlie the superficial, statistical, aspects that we can see.

When investigators obtain data, a favorite procedure is to fit a curve. Many investigators choose something simple like a straight line or a quadratic curve and fit it to dose response data by least squares or some similar procedure. These methods do have some advantages: They are simple and anyone can do it; because they are simple, they are transparent; that is, not only the investigator but his readers can see just what is going on. A problem, however, with such simple methods is that it is difficult, if not impossible, to take account of such factors as age at exposure, sex, time since exposure, and so on, that may be very important modifiers of risk. There is a second problem that often affects simple analyses: Very often no account is taken of the uncertainty that results from sampling variation. Error bars may be put on individual data points, but if a mathematical curve is fit to the data, there is often no real consideration of the uncertainty in that curve, both as to the particular parameter values and, in fact, even as to the functional form. The fact is that, as most of you appreciate, the induction of a cancer as a result of ionizing radiation exposure is a rare event. Let me recall to you that Kato & Schull, in their analysis of mortality among A-bomb survivors through the year 1978—more than 30 years after the bombings—estimated that there were fewer than 550 excess cancer deaths among nearly 300,000 A-bomb survivors in all Japan—and these 550 deaths were embedded in, and individually indistinguishable from, nearly 11,000 deaths from cancer overall.[2]

Estimated Total and Radiation Induced Cancer Deaths, 1950–1978 in 283,498 A-bomb Survivors in All Japan in 1950

	Mean Dose (T65) 16.1 rads	
	Excess Deaths	All Deaths
Leukemia	190.6	387
Other Cancer	335.7	10,421
TOTAL	526.3	10,808

Source: Kato & Schull, TR 12-80.

In fact, as you know, it is not possible to study deaths in *all* survivors. The actual data of Kato & Schull[2] were based upon just under 80,000 A-bomb survivors in the Life Span Study who had an average T65 dose of about 27 rads.

Estimated Total Radiation-Induced Cancer Deaths, 1950–1978 in 79,856 A-bomb Survivors in RERF Life Span Study

	Mean Dose (T65) 27.2 rads	
	Excess Deaths	All Deaths
Leukemia	90.6	180
Other Cancers	159.7	4,576
TOTAL	250.3	4,756

Source: Kato & Schull, TR 12-80

The estimated excess cancer deaths were only 250 out of 4,756 cancer deaths in total, or one in twenty, among 80,000 survivors, followed for 28 years. It is, therefore, not surprising, or it shouldn't be, that, given the very large statistical variability in the estimated excess, the uncertainties about the dosimetry and the need to postulate some mode or other of projecting observed, past risks into the future, that ideas about the total impact upon a population of a radiation exposure are rather vague.

Number of Excess Fatal Cancers Induced by 10 Rads in a Population of One Thousand During Its Remaining Lifetime

BEIR III Model	
Absolute Risk	
Linear-Quadratic	0.76
Linear	1.76
Relative Risk	
Linear-Quadratic	2.26
Linear	5.01

You are all familiar with the estimate that the lifetime risk of excess cancer in a normal population is about 1 or 2 per ten thousand per rad. The authors of the BEIR III report[3] fit linear and linear quadratic functions to the Hiroshima-Nagasaki data from 1950 to 1974. They allowed for a ten-year latent period for the cancers other than leukemia and projected forward in time according to two models for the time course of the induced solid cancers—whether an absolute, unchanging excess rate or an excess rate proportional to the natural rate which, of course, increases sharply with increasing age. The results you see in the table. The largest estimate is more than six times the smallest. And the remarkable thing is, that although the data set used by BEIR III was quite large—nearly four thousand cancer deaths—and an estimate of nearly 200 excess, radiation-induced cancer deaths, still the analysts were unable to distinguish among the models—they *all* fit the observed data within conventional statistical criteria!

We should note, further, that the variation shown here represents only that portion of the uncertainty that results from the choice of which model to use; the numbers are, in some sense, the "best" estimates that can be derived once the model is fixed. There is, however, within each model an uncertainty that results from simple, random statistical fluctuations. The BEIR III report does not present "standard deviations" applicable to these numbers but one judges that it is unlikely that the statistical uncertainties are smaller than fifty percent of the numbers shown. Thus, the actual uncertainty in the final result is of the order of a factor of ten.

We can note in passing that, given the very large uncertainties in the numbers for lifetime risk per rad, to say nothing of uncertainties in the doses, estimates of the number of excess deaths that will result in the Ukraine and Byelourussia over the next 70 years as a consequence of the accident at the Chernobyl power plant are so uncertain as to be almost worthless. The risk numbers are uncertain by a factor of the order of ten; how uncertain the dose estimates are I leave to someone else to say, but the result is that we simply do not know what are the likely consequences of Chernobyl for the populations of the surrounding areas.

Now that I've told you how bad things are, I owe it to you to give some hope for the future. Our estimates are going to improve very much in the next few years. The authors of the next BEIR report will have available to them data from a number of studies that were not available in the late 1970's when BEIR III was written. They will have the advantage of ten or more years of additional follow-up of the Japanese survivors.

Since a large fraction of the uncertainty in these numbers derives from different choices as to how to project to the future, every additional interval of follow-up reduces that component of the uncertainty and, ultimately, we, or our successors, will not have to make guesses about the future. In fact, the next report from the Life Span Study of the Radiation Effects Research Foundation will carry the data through the year 1985, forty years after the bombings. The addition of eleven years of follow-up information to the data available to the authors of the BEIR III report should help settle the question of which time projection model is appropriate for the cancers other than leukemia.

Finally, as you have been hearing, a new system of dosimetry has just been installed in Hiroshima and Nagasaki. Although the errors and uncertainties in dosimetry were not taken into account in the global uncertainties that I was talking about before, we hope that the new system will soon be accompanied by explicit estimates of uncertainty which should be, in any case, smaller than those of the old T65 system. In the new system the relative importance of neutrons is very much smaller than it was in the former T-65 system. This may make it easier to assess the risks from low-LET radiation but, on the other hand, it will probably be impossible to assess the risks from neutrons. On balance, perhaps, the gain outweighs the loss.

Why do we fit dose-response curves to data on carcinogenesis? For several different reasons:

- To smooth the observations.
- To suggest what the response might be at dose values not represented by data points. Those values may be within the range covered by the data—interpolation—or outside the data range—extrapolation.
- To enable exploration of the influence on risk of associated factors such as age at exposure, sex and others.
- To confront particular biological hypotheses about radiation carcinogenesis with data, to determine whether the hypotheses are, or are not, consistent with the real world.

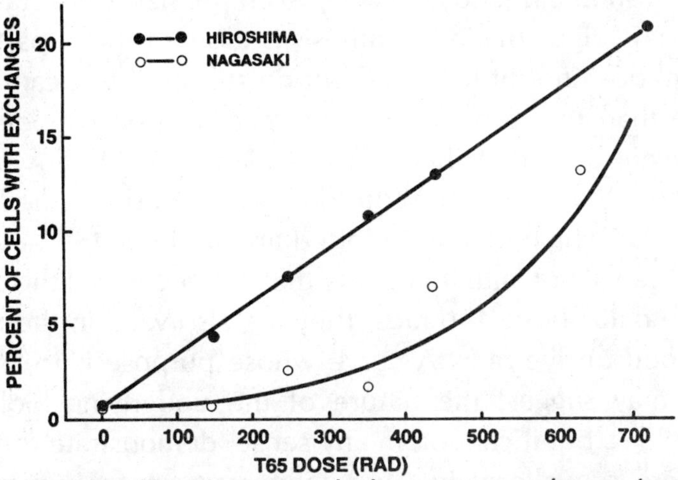

Fig. 1. Percent of lymphocytes with chromosome exchange aberrations in relation to estimated radiation exposure (kerma, T-65) in Hiroshima & Nagasaki atomic bomb survivors.

Figure 1 shows data published by Dr. Awa more than ten years ago comparing the chromosomal aberration response in Hiroshima and Nagasaki.[4] The doses used here are those from the old T-65 system and the data on the new system are somewhat different. The idea behind fitting curves is that we assume that there is *some* function that describes dose response; we don't know what it is, but we are confident that it is a smooth curve of some kind and, looking at the points, it seems reasonable to think that, in the case of Hiroshima, the response would be approximated well by a straight line and by a curve with a second degree term in the case of Nagasaki. No appeal to radiobiological theory is necessary to do this; such curves simply seem to describe the data reasonably well.

This is, clearly, an entirely appropriate procedure. But a problem does arise. Having fit the curves, and verified that the fits are good, there is a great tendency to forget that the curves were not chosen because of some theory concerning the underlying biological processes, a tendency to imagine that the curves do not merely smooth the data but represent *truth*. It should not be necessary to emphasize that a statistical test may tell us that a hypothesis is false, or that a particular model does not fit the data, but no statistical test can tell us more than that a hypothesis *may* be true or a model *may* represent a correct theoretical explanation. Moreover, you will note that, apart from the points at zero dose, the lowest doses for which data are plotted are at about 140 rads. It is simply an act of faith to assume that, however well the curves fit the data above 140 rads, they are also valid in the neighborhood of five rads. A curve whose purpose is to smooth data may *suggest* the nature of the underlying biological processes, but it cannot, in any sense, demonstrate them.

There is an alternative, and deeper, reason for curve fitting which serves quite a different purpose. That purpose is to hold up hypothetical biological models to the test of data. Theories about radiation carcinogenesis have implications concerning how the data should appear. Although curve fitting cannot tell us that a theoretical model is true, it may tell us that it is false.

Further, curve fitting may suggest biological models. A linear-quadratic dose response curve will suggest a two-hit model; although such a model is not necessarily entirely inconsistent with a linear response curve, a linear-quadratic response seems much more plausible.

Hattis,[5] in a recent paper, says of the present situation: " . . . at the present time much risk assessment takes the form of simple statistical curve fitting, without much more than lip service to the notion that the mathematical forms of the relationship fitted should bear some clear relationship to underlying causal mechanisms." And further: " . . . there is some tendency in the field to seize on the fact of incomplete knowledge . . . as an excuse to fit mathematical forms to data whose major virtue is computational simplicity." In my opinion, Hattis has described well the situation in our field. Until a few years ago we all fit linear, or linear-quadratic response functions, in the beginning by the method of least squares, and later, when computers became more widely available, by maximum likelihood. Just fifteen years ago Cox introduced the Proportional Hazards model. I will not try to describe the Cox model but content myself with saying that it has several very attractive features: It makes no assumptions whatever about the functional form of the underlying, baseline incidence or death rates; it provides a way of taking into account the effect of what are called "covariates"—age, sex and so on. Further, the covariation can be treated as varying in time as, for example, age at observation. The model is extremely flexible—almost too flexible: It is easy to introduce so many complex covariates that it may be possible to fit any data set almost perfectly while explaining very little.

There has been much confusion about terminology—what do we mean by absolute risk or relative risk, and what are the implications of these measures? A very useful set of definitions is contained in the draft UNSCEAR report, which distinguishes between measures that describe what the data tell us and concepts that describe how the world works.

A particular experience can be summarized in terms of the Relative Risk or the Excess Risk:

Hypothetical Data		
Number of Cancers Observed		43
Number Expected		27
Relative Risk	1.59	
Excess Risk	16 cases	
or 1.8 per 10,000 person-rads, or 6.0 per million person-year-rads.		

The Excess Risk, of course, would ordinarily be expressed as per ten thousand person-rads, or per million person-year-rads or on some other base. In any case, the point of creating such a summarizing number is that we wish to extrapolate the experience that has been observed to other experiences; from Japanese A-bomb survivors to the population around Chernobyl or from uranium miners to residents of houses on the Reading Prong. Taken by themselves, the numbers are simply descriptive of what has happened. When we attempt to extrapolate to a different population, however, several questions arise:

Is the effect of radiation multiplicative upon all the other causes of cancer in a population? If so, the Relative Risk is a more secure basis for extrapolation. Or,

Is the effect simply additive, a constant independent of other carcinogenic influences on the population? If so, the Excess Risk should be used for extrapolation.

Of course, it may be the case that radiation is additive with respect to some of the other relevant risk factors but multiplicative with respect to others. A case in point is attempting to extrapolate the radiation risk for lung cancer from males to females, say. Since the proportion of regular cigarette smokers is smaller among females than among males, just how to extrapolate will depend upon how the radiation risk works in combination with cigarette smoking. Fortunately, we have some direct experience to help guide us, from the A-Bomb survivors:

Risk of Fatal Lung Cancer, 1950–1978 RERF Life Span Study

	Males	Females
Number of fatal lung cancers, 100+ Rads	37	26
Relative Risk 100+ vs. 0 Rads	1.63	3.10
Excess Risk per Million P. R.	21.9	21.5

Given the relatively small numbers of lung cancer deaths in each sex group, it is almost surely fortuitous that the excess risks in the two high dose groups are almost identical, whereas the relative risks are almost twice as large in females as in males. The suggestion, then—and it is no more than that—is that the differences in baseline lung cancer rates between the sexes result from cigarette smoking, industrial exposures and perhaps other factors that apply much more to males than to females and that there is no interaction between acute radiation exposures such as those experienced in Japan and those factors. Other, more careful and sophisticated analyses of lung cancer in the A-Bomb survivors have been interpreted as showing that, *in that experience*, smoking and radiation appear to impose risks independently, that is to say, additively. I emphasize the phrase "in that experience" because, as you know, data concerning this issue, derived from the uranium miner studies, have pointed to the opposite conclusion. The explanation for this discrepancy may lie in the very different conditions of exposure—within a few seconds in one case, and protracted over many years in the other.

So much for relative risk and excess risk. We turn to two possible "time-projection models". No epidemiological study has yet followed a large, irradiated population to extinction. Even the study in Japan, which now extends to forty years, is barely more than half completed; more than 60 percent of the cohort survived through the year 1985 and it will be another 30 to 40 years before the experience is essentially complete. But, naturally enough, people want to know what the future holds. Many of us have been caught up in the game called "How many excess cancer deaths will ultimately result in the population around Chernobyl in consequence of the accident there?" To do this in a sensible way one must somehow project a partial experience into the future. For leukemia there is no such problem—it appears that the wave of excess leukemia cases that results from a radiation exposure is essentially complete by about 30 years later. One can, then, estimate the total number of excess cancers by assuming that one knows the ratio between the leukemias and other cancers or one can examine the data to find evidence of the time-pattern of the excess cancers, and project forward. One can use an Absolute Risk projection, that is, one in which the excess risk per year per rad is simply added to the baseline, naturally occurring risk, or one can use a Multiplicative model, sometimes called a Relative Risk projection model, in which it is assumed that the excess risk in a cohort increases proportionally with the increase in baseline rates that accompanies aging. The authors of BEIR III used both types of projection model with very different results, as we have seen. At this time, many of us have opted for the multiplicative time projection model, or relative risk projection model, if one prefers, mostly on the grounds that the constant absolute risk model is clearly not applicable to the experience in Japan.

Up to now, I've been talking about problems in choosing models, assessing uncertainty, deciding how to project from the past to the future and, at least implicitly, how to project from levels of dose and dose rate at which we can measure effects to levels that most concern us—a few rads, received over a rather long time. The problems that we have been discussing are mostly concerned with what are generally considered to be the area of radiobiology and biostatistics. There are also problems that are more strictly epidemiological.

Let me start out by saying that there is no real, theoretical foundation for the use of statistical methods in observational studies on humans. We all use them, however, including myself, so what am I talking about?

R. A. Fisher, whose name you will recognize as that of the father of the subject of "experimental design" more than fifty years ago, in his book, "The Design of Experiments" wrote

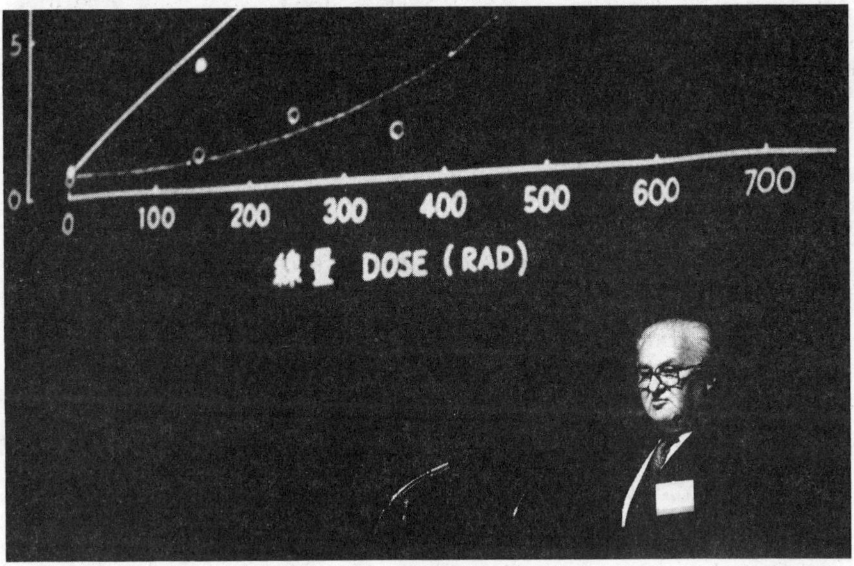

Seymour Jablon

" . . . the physical act of randomization . . . is necessary for the validity of any test of significance."[6]

It is possible to physically randomize mice, say, in a laboratory experiment. It is seldom possible to randomize people, as in Hiroshima or in a uranium mine or to those receiving radiotherapy for ankylosing spondylitis. Since we do not, in fact cannot, assign persons at random to one or another exposure group we take so-called "natural experiments" as we find them—we have no choice. But when we subsequently compare some outcome between two groups—say the occurrence of cancer, the fact is that any difference that is found *may* be a consequence of the difference in exposure to radiation or it may be a consequence of the fact that the two groups were different to begin with; differences in outcome might have been observed even if there had been no radiation exposure to anyone. Or, of course, any observed result may be a mixture, in unknown proportions, of the consequences of the two sources of difference: the treatment effect that we are trying to measure and the uncontrollable effect of underlying cohort differences.

As an extreme example, an investigator might compare lung cancer rates in residents of Salt Lake City with rates in Washington, D. C., in a search for the possible effect of the higher levels of background radiation in Salt Lake City. We know very well that the life-style that characterizes Mormons has a much greater influence on cancer rates than could conceivably result from differences in dose from background radiation of the order of 50 or 100 millirem yearly. Although few investigators would be so naive as to make such a comparison, the same problem, in principle, affects most epidemiological studies.

For example, beginning in the late 1950's a series of studies have examined childhood cancer rates in relation to whether the mother was exposed to abdominal x-rays during pregnancy. Without challenging the reported association between exposure of the fetus to x-rays and subsequent leukemia or other cancers, there are other differences than simply the x-ray exposure: first-born children were more likely to have been x-rayed whether they subsequently had cancer or not, and first-born children were also more likely to have cancer during the first ten years of life, whether they were x-rayed or not.[7]

The epidemiologic study of the Japanese A-bomb survivors is the most important source we have for estimating the risks of exposure to ionizing radiation upon subsequent health. The data have customarily been treated on the assumption that survivors of a given sex and age at exposure were essentially identical, that differences in outcome associated with estimated radiation dose are, statistical variability aside, the direct result of the radiation dose. This may not be true. Japanese cities are not characterized by very sharp socio-economic differences between different districts, as many cities are in the United States. Japanese cities are not, however, uniform as to the character of those who live or work in particular areas. Districts where radiation doses were low or high were not entirely comparable with respect to the character of the residents.[8]

Peterson and his colleagues examined cancer mortality among A-bomb survivors after dividing each city into eight wedge-shaped octants, centered on the hypocenter.[9] They were interested in the question whether radiation doses were independent of direction, as well as distance, from the hypocenters, postulating that differences might have arisen because of asymmetry in the prompt radiation output or because of local fallout in particular areas. In fact, they found some significant differences in cancer rates among the octants. On consideration, however, they concluded that the most likely explanation of the differences was to be found in what they called "risk factors other than dose asymmetry" which I conceive to be an elaborate way of referring to the fact that persons who were in different parts of the cities at the time of the explosions had not been randomized to these locations; they were in particular places for their own reasons: they worked there or lived there or happened to be passing through. And the conclusion has to be that we have no good reason to believe that, apart from radiation exposure, they would all have been subject to the same age and sex specific cancer rates. No less than in the United States, cancer death rates in Japan vary from city to city. There is no reason to suppose that within a particular city rates will be uniform.

There are two more sources of error that are usually ignored although they may affect conclusions in important ways. I refer to errors in specification of the radiation dose to individuals, and to error in specification of the outcome observations that is, to errors in diagnosis. Most mortality studies rely on death certificates to learn about causes of death and studies in Japan,[10] Great Britain[11] and the United States[12] have concluded alike that the cause of death, as stated on a death certificate, is often incorrect. If one is satisfied to know that a death did or did not result from a cancer, any cancer, certificates are correct about 90 percent of the time. If, however, one is interested in specific cancers, for some cancer sites the diagnostic accuracy is very poor: In the Japanese series only about half of the respiratory system cancers found at autopsy were detected if reliance was placed exclusively on death certificate notifications.

The other variable essential to an assessment of risk per unit dose is, of course, the radiation dose or exposure to individuals. Unlike the experimenter, the epidemiologist is normally in the position of attempting to reconstruct the radiation dose retrospectively. There are inevitably errors in assessing doses to individuals; even if the dosimetry *system* is correct, it is usually impossible to attain to perfect accuracy in establishing the precise parameters of exposure to each person. This is equally true of Japanese atomic bomb survivors and of underground miners. If, as seems often to be the case, the errors in individual dose estimates are distributed lognormally, there are some unpleasant consequences:[13] The mean dose in a group will be overestimated and, therefore, the estimate of risk per unit dose will be underestimated; because the relative importance of the dosimetry errors will be much larger for large estimated doses, the net result will be a tendency towards obtaining a concave dose response curve. This will be of the same form as the concavity that is usually attributed to cell sterilization at the larger doses but how to apportion the responsibility between these two causes for a concave dose-response curve will usually be a puzzle.

How then do we find truth, finally? The answer, I believe, lies in the search for consistency among studies. Every study is subject to bias and no single study can be relied on exclusively for the reasons just given. We can, however, hope that different studies, done in differing circumstances, will be subject to bias in different ways. To the extent that we can find consistency in studies done in widely varying contexts we may be justified in believing that we have caught hold of a piece of scientific truth. It may be true in general that a healthy skepticism is an epidemiologist's best friend, but it is certainly true of epidemiological studies of radiation effects. Any new scientific finding requires independent verification. What we need in radiation epidemiology is verification squared.

References

1. J. H. Lawrence. Quoted in *A Random Walk in Science*, compiled by R. L. Weber (Crane, Russak & Co., New York, 1973.)
2. H. Kato and W. J. Schull. "Studies of the Mortality of A-Bomb Survivors 7. Mortality, 1950–1978: Part I. Cancer Mortality," Radiat. Res. *90*, 395 (1982).
3. BEIR. Biological Effects of Ionizing Radiation. *The Effects on Populations of Exposure to Low Levels of Ionizing Radiation.* (National Research Council-National Academy of Sciences, Washington, DC, 1980).
4. A. A. Awa. "Chromosome Aberration in Somatic Cells," J. Radiat. Res. *16*, Suppl., 122 (1975).
5. D. B. Hattis. "The Promise of Molecular Epidemiology for Quantitative Risk Assessment," Risk Analysis 6, 181 (1986).
6. R. A. Fisher. *The Design of Experiments.* Hafner-Publishing Company, New York, (1947).
7. A. Stewart, J. Webb and D. Hewitt. "A Survey of Childhood Malignancies," Brit. Med. J. *1*, 1495, (1958).
8. S. Jablon, M. Ishida and M. Yamasaki. "JNIH-ABCC Life Span Study, Hiroshima-Nagasaki, Report 3, Mortality October 1950–September 1960," ABCC Technical Report 15-63 (1963).
9. A. V. Peterson, Jr., R. L. Prentice, T. Ishimaru, H. Kato and M. Mason. "Investigation of Circular Asymmetry in Cancer Mortality of Hiroshima and Nagasaki A-Bomb Survivors," Radiat. Res. *93*, 1984 (1983).
10. S. Jablon, D. M. Angevine, Y. S. Matsumoto and M. Ishida. "On the Significance of Cause of Death Certificates in Hiroshima & Nagasaki, Japan," National Cancer Institute Monograph 19, *Epidemiological Approaches to the Study of Cancer and Other Chronic Diseases* (National Cancer Institute, Bethesda, 1966).
11. M. A. Heasman and L. Lipworth. *Accuracy of Certification of Cause of Death,* Studies on Medical and Population Subjects No. 20 (Her Majesty's Stationery Office, London, 1966).
12. C. Percy, E. Stanke, III and L. Gloeckler.: "Accuracy of Cancer Death Certificates and its Effect on Cancer Mortality Statistics," Am. J. Pub. Health *71*, 242 (1981).
13. E. S. Gilbert. "Some Effects on Random Dose Measurement Errors on Analyses of Atomic Bomb Survivor Data," Radiat. Res. *98*, 591 (1984).

Warren K. Sinclair, President of the NCRP presenting Commemorative plaque to the Lecturer

NATIONAL COUNCIL ON RADIATION PROTECTION AND MEASUREMENTS
LAURISTON SALE TAYLOR
Presented to
SEYMOUR JABLON
by the
NATIONAL COUNCIL ON RADIATION PROTECTION AND MEASUREMENTS
in commemoration of his presentation of
The Eleventh Annual Lauriston S. Taylor Lecture
HOW TO BE QUANTITATIVE ABOUT RADIATION RISK ESTIMATES
on April 8, 1987
WARREN K. SINCLAIR
President
NCRP

Scientific Session

The Impact of the New Dosimetry (continued)

Itsuzo Shigematsu,
Chairman

Chromosome Aberration Data For A-Bomb Dosimetry Reassessment

Akio A. Awa [1], Kazuo Ohtaki [1], Massahiro Itoh [2], Takeo Honda [2], Dale L. Preston [3], and Mary E. McConney [3]
Department of Genetics [1]
Department of Radiobiology (Nagasaki) [2]
and Research Information Center [3]
Radiation Effects Research Foundation

Summary

Cytogenetic data, derived from cultured lymphocytes of A-bomb survivors and controls in the ABCC-RERF Adult Health Study cohort, have been analyzed to determine differences in the dose-response relationships for chromosome aberrations between the T65D and DS86 dose estimates and to assess differences between Hiroshima and Nagasaki. The data consists of blood samples collected between 1968 and 1980 from 1245 individuals, 788 in Hiroshima and 457 in Nagasaki, who have a DS86 kerma estimate of less than 4 Gy (678 in Hiroshima and 381 in Nagasaki) or who were not in the cities at the time of the bomb (110 in Hiroshima and 76 in Nagasaki). The data analysis is based on the number of cells in which at least one structural chromosome aberration was identified as a proportion of the number of cells examined (generally 100 cells per case). The dose response is examined by looking at total tissue kerma in air at the survivor location adjusted for the effects of shielding by structures or terrain and absorbed dose to the active bone marrow for both T65D and DS86 in each city. Binominal regression models were used to describe the relationship between the proportion of aberrant cells and kerma or marrow dose. The variability of the observed number of aberrant cells as a function of estimated dose is much greater than one would expect under a binomial model in which the aberration rate is a simple linear or quadratic function of dose, consequently, all models have been adjusted for extra-binomial variability.

For a linear dose response model, the average percentage of cells with chromosome aberrations increases less rapidly with dose in Nagasaki than in Hiroshima. The magnitude of the city difference in slopes is less for DS86 than for T65D, though the difference is statistically significant for both kerma and bone marrow dose with either dosimetry. The DS86 slope for kerma estimates is about 60% larger than the corresponding T65D slope. For marrow dose, however, there is very little change between the two dosimetries; moreover, this change is even smaller in Hiroshima than in Nagasaki. Further analyses to test nonlinearity in the dose-response function show that there appear to be significant departures ($p<.001$) from linearity with both DS86 and T65D dosimetries for both kerma and bone marrow dose.

Introduction

It is well established that the frequency of the in vivo induction of chromosomal aberrations in the peripheral blood lymphocytes responds sensitively to exposure to ionizing radiation [1]. Thus, one expects to see a dose-dependent increase in the frequency of cells with induced chromosomal aberrations with increasing radiation dose among the survivors of the atomic bombings of Hiroshima and Nagasaki [2,3]. Cytogenetic data on the A-bomb survivors have been used previously as a reliable biological endpoint for the A-bomb dosimetry reassessment [3,4,5,6].

The purpose of this report is to describe the difference in the dose-response relationships for chromosome aberration frequencies of A-bomb survivors between the T65D and DS86 dose estimates as well as to determine whether there is a difference between Hiroshima and Nagasaki. Since A-bomb survivors received acute whole-body exposure, chromosomal aberrations that are observable now in T lymphocytes were induced at the time of the bombing not only in mature lymphocytes in the circulating blood but also in other lymphocyte-producing organs, such as bone marrow, spleen, thymus and lymph nodes. Thus, the analysis of the dose-response relationship for chromosome aberrations is presented using both tissue kerma and the absorbed dose for the active bone marrow.

Materials and Methods

Cytogenetic Methods

The data consists of usable blood samples collected between 1968 and 1980 from survivors and controls in the Adult Health Study. Usable blood samples are those from which a successful culture with at least 30 scorable cells was made. After excluding individuals who had received radiotherapy or radioisotope treatment at any time in the past, blood samples from 1,491 individuals are available for analysis. For some of these individuals there are more than one usable blood sample, however, in this paper, we only consider the most recent observation for each person.

These blood samples were cultured for 52 hours and then cells were harvested for chromosome preparations using the conventional Giemsa stain procedure as described elsewhere [3,7]. The culture time was chosen by assuming that most of the observed metaphases would be in their first in vitro cell division at 52 hours however, using the FPG method for sister-chromatid differential stain [8], it was discovered that in fact, the proportion of cells in their second vitro cell division varied from 0 to 40% of the total metaphases analysed (Ohtaki *et al.*, personal communication).

All slides were coded and microscopic examinations were performed without knowledge of an individual's exposure status. The chromosomes were classified into seven chromosome groups (A–G) directly under the microscope. Except for gross chromosome aberrations that are easily recognizable, structural abnormalities of the stable type were suspected when loss or gain of a chromosome(s) in any of these chromosome groups was identified in the metaphases examined. These cells were photographed for further detailed karyotype analysis and structural chromosome aberrations were classified according to the ISCN (1985) [9]. Specifically, stable aberrations include reciprocal translocations, pericentric inversions and deletions without fragments. For each sample, the cytogenetic data used in these analyses are the total number of cells examined (usually 100 metaphases per sample) and the number of cells in which at least one stable aberration was observed. No attempt has been made to use information on unstable aberrations or on the total number of aberrations seen.

The Dose Data

Table 1 summarizes the distribution of people for whom usable blood samples are available by city and exposure status and type of dose estimate. Among the 1,491 samples, 1,305 are from survivors for whom T65D estimates are available while 186 were obtained from people who were not in the cities (NIC) at the time of the bombings (ATB). As discussed by Preston and Pierce [10], DS86 estimates cannot be computed for all survivors with T65D estimates. The total number of usable blood samples from exposed survivors for whom DS86 estimates are available is 1,101.

In order to describe the impact of the change in dosimetry for the survivors included in the analysis and to provide more information on the exposure status of survivors for whom DS86 estimates are not available, Table 2 contains the city specific joint distributions over T65D and DS86 exposure kerma estimates are generally smaller than

TABLE 1—*Number of usable blood samples by exposure status and city. Information on the distribution for samples included in these analyses are indicated in the top part of the table. The data in the bottom part of the table are for samples which were excluded from the main analyses of this paper for reasons discussed in the text.*

Exposure Status	City		
	Hiroshima	Nagasaki	Total
Included in analyses:			
Not in city	110	76	186
DS86 < 4Gy	678	381	1,059
Subtotal	788	457	1,245
Excluded from analyses:			
DS86 $\geq$ 4Gy	33	9	42
T65D only	33	171	204
Subtotal	66	180	246
Total	854	637	1,491

the T65D estimates. It is also apparent that the proportion with T65D kerma estimates greater than 5 mGy for whom DS86 estimates cannot be computed is higher in Nagasaki than in Hiroshima. These patterns are similar to those noted for the Life Span Study as a whole [10]. An important reason for the heavier losses among Nagasaki survivors with relatively high doses is that the DS86 system does not currently include procedures for the estimation of doses for people exposed in factories or who were shielded by terrain.

All analysis is based on the 1,245 individuals who have a DS86 kerma estimate less than 4 gray (1,059 people) or who were not in the cities at the time of the bombs (186 people). Because of the relationship between T65D and DS86 kerma estimates, this is roughly equivalent to the exclusion of survivors with T65D kerma estimates in excess of 6 gray. These high dose cases were excluded because, given that these people survived both the actual blast and the radiation, it is unlikely that their true dose is substantially smaller than the estimated dose [10].

We are interested in relating the proportion of cells with aberrations to some measure of radiation exposure. It is likely that most of the cells observed in this study originated from cells which were in lymphocyte-producing organs, such as bone marrow, spleen, thymus and lymph nodes, as well as in the circulating blood. Because of this the primary emphasis here is placed on absorbed dose to the active bone marrow. The T65D system did not include procedures for estimation of absorbed dose to specific organs. Here, we use Kerr's [11] estimated transmission factors for bone marrow applied to the T65D kerma estimates to get the absorbed dose estimate. DS86 provides absorbed marrow estimates for individual survivors. Table 3 contains Kerr's average marrow dose factors and information about the DS86 marrow dose factors for the survivors used in these analyses.

Earlier analyses of these data have generally been based on estimates of total tissue kerma in air at the survivor location adjusted for the effects of shielding by structures or terrain. To be comparable with the earlier work, we also analyse the aberration rate as a function of kerma.

Statistical Methods

Binomial regression models are appropriate to describe the relationship between the proportion of aberrant cells and kerma or dose. One of

TABLE 2—*Hiroshima*

T65DR Kerma in Gray	DS86 Kerma in Gray: NIC	0–.005	.005–.10	.10–.50	.50–1.00	1.00–2.00	2.00–3.00	3.00–4.00	Subcohort total	DS86 > 4.Gy	Unknown DS86	T65DR Kerma total	% of T65DR w/unk. DS86 Kerma
NIC	110								110	0	0	110	0.0
0.–.005		252	1						253	0	0	253	0.0
.005–.10			12						12	0	8	20	40.0
.10–.50				51	5				56	0	4	60	6.7
.50–1.00				9	18	2			29	0	2	31	6.5
1.00–2.00					67	80			147	0	9	156	5.8
2.00–3.00						57	25	1	83	0	5	88	5.7
3.00–4.00						11	32	8	51	0	1	52	1.9
4.00+						2	26	19	47	33	4	84	4.8
DS86 Kerma Total	110	252	13	60	90	152	83	28	788	33	33	854	3.9

Nagasaki

T65DR Kerma in Gray	DS86 Kerma in Gray: NIC	0–.005	.005–.10	.10–.50	.50–1.00	1.00–2.00	2.00–3.00	3.00–4.00	Subcohort total	DS86 > 4.Gy	Unknown DS86	T65DR Kerma total	% of T65DR w/unk. DS86 Kerma
NIC	76								76	0	0	76	0.0
0–.005		173							173	0	5	178	2.8
.005–.10		2	1						3	0	0	3	0.0
.10–.50			18	16					34	0	19	53	35.8
.50–1.00				20	4				24	0	19	43	44.2
1.00–2.00				12	33	4			49	0	63	112	56.2
2.00–3.00					11	32			43	0	43	86	50.0
3.00–4.00					1	19	2	1	23	0	11	34	32.4
4.00+						6	21	5	32	9	11	52	21.2
DS86 Kerma Total	76	175	19	48	49	61	23	6	457	9	171	637	26.8

TABLE 3—*DS86 and T65D marrow dose transmission factors. The values shown for DS86 are mean values computed from the marrow dose to kerma ratios for the survivors included in these analyses. N-gamma refers to neutron capture gamma rays generated as neutrons interact with the marrow or surrounding tissues.*

	Dosimetry	
Radiation	T65D	DS86
gamma	0.56	0.79
neutron	0.26	0.33
n-gamma	0.07	0.57

the striking features of these data, which has been noted by many authors [3,12,13] is that the variability of the observed number of aberrations as a function of estimated dose appears to be much greater than one would expect under a binomial model in which the aberration rate was a simple linear or quadradic function of dose. Failure to take this extra-binomial variation into account leads to underestimates of the variance of the parameter estimates and invalidates significance tests. In order to adjust for this over dispersion we have assumed that the extra-variability over the range of interest here increases with the square of the true proportion of cells with aberrations. This model does not differ appreciably from the beta-binomial model used by Otake and others [12,13] in analysis of these data. All models have been adjusted for the extra-binomial variation using the method of moments suggested by Moore [14] and all models are fit using iteratively reweighted least squares with the weight at each iteration determined by the variance function [15].

Results

Table 4 contains the information on the distribution of the number of samples, cells counted, and cells with aberrations by dose (kerma) category for each city. In addition, the mean aberration rate per person (in percent) and the standard deviation of the rate are given for each category. Since the number of cells counted per person is not constant the mean aberration rate is not equal to the ratio of the number of cells with aberrations to the total number of cells counted in a dose category

TABLE 4a—*Summary of Chromosome Aberration Data by DS86 Kerma (Gy) and City.* Only samples for individuals with total DS86 Kerma < 4Gy are included. The mean and standard deviation of the percent aberrations in each dose category are computed from the results for each usable sample.

Kerma group	Mean Kerma	Usable blood Samples	Total cells counted	Total Aberrant cells	Mean % Aberrations	Standard deviation % Aberrations
Hiroshima						
NIC	0.0	110	9,968	130	1.24	1.40
0-.005	0.0	252	24,271	233	.96	1.30
.005-.10	.048	13	1,188	13	1.09	1.24
.10-.50	.295	60	5,624	105	1.82	1.74
.50-1.0	.779	90	8,697	436	4.98	4.08
1.0-2.0	1.43	152	14,500	1,309	9.00	8.09
2.0-3.0	2.45	83	8,057	1,340	16.67	11.09
3.0-4.0	3.37	28	2,643	577	21.58	12.15
Summary						
NIC-.005	0.0	362	34,239	363	1.04	1.33
.005-4.0	1.42	426	40,709	3,780	9.22	9.74
Nagasaki						
NIC	0.0	76	7,522	106	1.40	1.26
0-.005	0.0	175	17,017	239	1.39	1.50
.005-.10	.070	19	1,853	22	1.16	1.34
.10-.50	.299	48	4,584	83	1.75	2.68
.50-1.0	.738	49	4,768	140	2.87	2.91
1.0-2.0	1.42	61	5,990	433	7.42	7.90
2.0-3.0	2.39	23	2,215	369	16.25	12.70
3.0-4.0	3.45	6	600	111	18.50	14.12
Summary						
NIC-.005	0.0	251	24,539	345	1.39	1.43
.005-4.0	1.04	206	20,010	1,158	5.75	8.36

(simple mean rate). The mean aberration per person is a more appropriate summary of the data than the simple mean rate.

Linear Models

First we look at linear models to assess the impact of changes in dosimetry on the aberration rate dose-response and to assess the inter-city differences in the response. Table 5 contains parameter estimates and standard errors for city-specific linear response models of the form

$$E(p;d) = b_{Oc} + b_{ic}\underline{d}$$

Where $\underline{d}$ denotes either the kerma or marrow doses and the b_{ic} are city-specified regression parameters. Results are given for both kerma and

TABLE 4b—*Summary of Chromosome Aberration Data by DS86 Marrow Dose (Gy) and City.*

Only samples for individuals with total DS86 Kerma < 4Gy are included. The mean and standard deviation of the percent aberrations in each dose category are computed from the results for each usable sample.

Dose group	Mean Dose	Usable blood Samples	Total cells counted	Total Aberrant cells	Mean % Aberrations	Standard deviation % Aberrations
Hiroshima						
NIC	0.0	110	9,968	130	1.24	1.40
0-.005	0.0	252	24,271	233	.96	1.30
.005-.10	.055	19	1,788	22	1.22	1.17
.10-.50	.307	76	7,116	188	2.60	3.24
.50-.1.0	.749	126	12,087	715	5.83	4.38
1.0-1.5	1.24	87	8,348	842	10.14	9.20
1.5-2.0	1.73	63	6,170	1,004	16.26	10.67
2.0+	2.40	55	5,200	1,009	19.35	12.10
Summary						
NIC-.005	0.0	362	34,239	363	1.04	1.33
.005-2.0+	1.10	426	40,709	3,780	9.22	9.74
Nagasaki						
NIC	0.0	76	7,522	106	1.40	1.26
0-.005	0.0	175	17,017	239	1.39	1.50
.005-.10	.064	24	2,303	27	1.13	1.26
.10-.50	.312	58	5,634	106	1.84	2.57
.50-1.0	.748	56	5,359	255	4.92	6.37
1.0-1.5	1.23	35	3,499	266	7.60	7.42
1.5-2.0	1.76	23	2,263	397	15.29	12.75
2.0+	2.62	10	952	157	15.70	13.29
Summary						
NIC-.005	0.0	251	24,539	345	1.39	1.43
.005-2.0+	.83	205	20,010	1,158	5.75	8.36

marrow dose for both dosimetries in Table 5. The last column contains p-values for tests of the significance of inter-city differences in the intercepts and slopes.

The parameter estimates in this and all subsequent tables are given in terms of the percentage of cells with aberrations which is 100 times the aberration rate. Thus, for example, an intercept of 1.0 means that aberrations are seen in about 1% of the cells of nonexposed or zero dose survivors. While an estimated slope of 6 means that the expected percentage of cells with aberrations increased by about 6 per gray.

Looking first at the intercepts, there appears to be a small, but statistically significant difference in the background rates for the two

TABLE 4c—*Summary of Chromosome Aberration Data by T65D Kerma (Gy) and City*

Only samples for individuals with total DS86 Kerma < 4Gy are included. The mean and standard deviation of the percent aberrations in each dose category are computed from the results for each usable sample.

Kerma group	Mean Kerma	Usable blood Samples	Total cells counted	Total Aberrant cells	Mean % Abberrations	Standard deviation % Aberrations
Hiroshima						
NIC	0.0	110	9,968	130	1.24	1.40
0-.005	0.0	253	24,371	235	.96	1.30
.005-.10	.048	12	1,088	11	1.01	1.26
.10-.50	.299	56	5,274	93	1.74	1.50
.50-1.0	.706	29	2,677	123	4.36	4.13
1.0-2.0	1.45	147	14,118	802	5.73	4.44
2.0-3.0	2.49	83	7,946	885	11.16	7.76
3.0-4.0	3.46	51	5,053	933	18.69	13.06
4.0+	4.69	47	4,453	931	20.57	11.34
Summary						
NIC-.005	0.0	363	34,339	365	1.04	1.33
.005-4.0+	2.01	425	40,609	3,778	9.24	9.74
Nagasaki						
NIC	0.0	76	7,522	106	1.40	1.26
0-.005	.002	173	16,888	239	1.39	1.50
.005-.10	.015	3	229	0	0	0
.10-.50	.268	34	3,258	35	1.03	1.22
.50-1.0	.754	24	2,326	46	1.92	2.22
1.0-2.0	1.45	49	4,725	143	2.95	3.48
2.0-3.0	2.51	43	4,240	221	5.16	4.36
3.0-4.0	3.50	23	2,248	185	8.04	9.41
4.0+	4.71	32	3,113	528	17.24	12.63
Summary						
NIC-.005	.001	249	24,410	345	1.40	1.43
.005-4.0+	2.10	208	20,139	1,158	5.69	8.34

cities. Specifically, the background aberration rate in Nagasaki is estimated to be slightly greater than that in Hiroshima. In order to gain additional insight into this question a simple analysis was carried out using only data on the NIC group and persons with DS86 kerma estimates of less than 5 mGy. The city-specific rate estimates for these groups are shown in Table 6. Based upon these estimates, it appears that the differences are due almost entirely to the low frequency of aberrations in Hiroshima survivors with DS86 kerma less than 5 mGy. There is virtually no difference between the aberration frequency for members of the Hiroshima NIC group, the Nagasaki NIC group and the Nagasaki group with DS86 kerma less than 5 mGy.

TABLE 4d—*Summary of Chromosome Aberration Data by T65D Marrow Dose (Gy) and City*

Only samples for individuals with total DS86 Kerma < 4Gy are included. The mean and standard deviation of the percent aberrations in each dose category are computed from the results for each usable sample.

Dose group	Mean Dose	Usable blood Samples	Total cells counted	Total Aberrant cells	Mean % Abberrations	Standard deviation % Aberrations
Hiroshima						
NIC	0.0	110	9,968	130	1.24	1.40
0-.005	0.0	253	24,371	235	.96	1.30
.005-.10	.060	32	3,058	38	1.24	1.22
.10-.50	.272	63	5,781	175	2.89	3.11
.50-1.0	.738	134	12,898	713	5.62	4.53
1.0-1.5	1.24	87	8,292	841	10.15	6.89
1.5-2.0	1.73	54	5,327	942	17.83	12.89
2.0+	2.41	55	5,253	1,069	20.09	11.46
Summary						
NIC-.005	0.0	263	34,339	365	1.05	1.33
.005-2.0+	1.06	425	40,609	3,778	9.24	9.74
Nagasaki						
NIC	0.0	76	7,522	106	1.40	1.26
0-.005	.001	175	17,017	239	1.39	1.50
.005-.10	.086	9	853	12	1.33	1.73
.10-.50	.260	45	4,331	61	1.36	1.82
.50-1.0	.737	46	4,456	110	2.41	3.22
1.0-1.5	1.26	38	3,709	175	4.64	4.29
1.5-2.0	1.71	27	2,652	212	7.85	7.77
2.0+	2.52	41	4,009	588	14.92	12.51
Summary						
NIC-.005	0.0	251	24,539	345	1.39	1.43
.005-2.0+	1.19	206	20,010	1,158	5.75	8.36

The slope estimates shown in Table 5 demonstrate that the average percentage of cells with chromosome aberrations increases less rapidly with dose in Nagasaki than in Hiroshima. The magnitude of the city difference in slopes is less for DS86 than for T65D; however, this difference is statistically significant for both kerma and marrow dose with either dosimetry. Table 7 summarizes the relative change in the slope of the aberration response function for the new dosimetry. Separate results are given for each city. and for the average of both cities combined. In terms of kerma estimates the DS86 slope is about 60% larger than the corresponding T65D estimate for these data, while for marrow dose there is very little change. The change in the Hiroshima slopes are less marked than the change for Nagasaki.

TABLE 5—*Regression coefficients and significance tests in linear models of the chromosome aberration rate (in percent). Parameter estimates and intercity significance tests are given for kerma and marrow dose risks with both the DS86 and T65D dosimetries. Separate linear models have been fit for each city. Estimates were computed using quasi-likelihood models which allow for overdispersion. The intercept is the percentage of cells with aberrations among persons whose estimated dose is 0. The slope indicates the change in the percentage of cells with aberrations per Gy.*

Dosimetry	Term	Regression Coefficients				p-Value for Inter-City Comparison
		Hiroshima		Nagasaki		
		Coeff.	Std.Err.	Coeff.	Std.Err.	
DS86 kerma	Intercept	1.01	0.07	1.30	0.09	.09
	Slope	5.6	0.3	4.0	0.4	<.001
T65D kerma	Intercept	1.03	0.07	1.28	0.9	.02
	Slope	3.9	0.2	1.9	0.2	<.001
DS86 marrow dose	Intercept	1.01	0.07	1.26	0.9	.01
	Slope	7.2	0.4	5.1	0.4	<.001
T65D marrow dose	Intercept	1.03	0.07	1.28	0.9	.02
	Slope	7.7	0.4	3.5	0.3	<.001

TABLE 6—*Change in slope relative to T65D slope by city and for both cities combined.*

Comparison	Measure of Exposure	
	Kerma	Marrow Dose
Hiroshima	44%	-6%
Nagasaki	111%	46%
Average for both cities	58%	7%

TABLE 7—*Comparison of background aberration rates by city and exposure status.*

Exposure Status	City	
	Hiroshima	Nagasaki
<5 mGY DS86 kerma	.96 (.07)	1.3 (0.1)
Not-in-city	1.29 (∼.08)	1.4 (∼.1)

Non-linearity of the dose response function

In order to investigate nonlinearity in the dose-response function we used city specific linear quadratic models for the aberration rate:

$$E(p;d) = b_{0c} + b_{ic} + b_{2c}d^2.$$

Using the asymptotic significance level of the quadratic term in a model for both cities combined as a rough guide to the statistical significance of the nonlinearity in these models, there appear to be significant departures ($p<.001$) from linearity with both the DS86 and T65D dosimetries for both kerma and marrow dose risks. Table 8 contains parameter and standard error estimates for the city-specific modes. The significance tests for the intercity comparisons (not shown) parallel those for the linear model: there are significant differences in both the intercepts and the dose response.

The plots in Figure 1 summarize the relationship between the aberration rate and kerma or marrow dose estimates for both dosimetries. The points shown are the unweighted average frequencies of cells with chromosome aberrations in each dose group plotted at the city specific average doses (as reported in Table 4). The linear and linear quadratic fitted curves are superimposed.

TABLE 8—*Regression coefficients and significance tests in quadratic models of the chromosome aberration rate (in percent). Parameter estimates and intercity significance tests are given for kerma and marrow dose risks with both the DS86 and T65D dosimetries. Separate linear models have been fit for each city. Estimates were computed using quasi-likelihood methods which allow for overdispersion. The intercept is the percentage of cells with aberrations among persons whose estimated dose is 0.*

		Regression Coefficients			
Dosimetry	Term	Hiroshima		Nagasaki	
		Coeff.	Std.Err.	Coeff.	Std.Err.
DS86 kerma	Intercept	1.04	0.07	1.4	0.1
	Linear	3.9	0.6	0.5	0.7
	Quadratic	1.0	0.3	2.3	0.5
T65D kerma	Intercept	1.05	0.07	1.4	0.09
	Linear	3.0	0.4	-0.1	0.3
	Quadratic	0.4	0.2	0.7	0.1
DS86 marrow dose	Intercept	1.04	0.07	1.4	0.1
	Linear	5.4	0.8	1.1	0.9
	Quadratic	1.4	0.5	3.3	0.8
T65D marrow dose	Intercept	1.05	0.07	1.4	0.09
	Linear	5.8	0.8	-0.2	0.6
	Quadratic	1.5	0.6	2.2	0.4

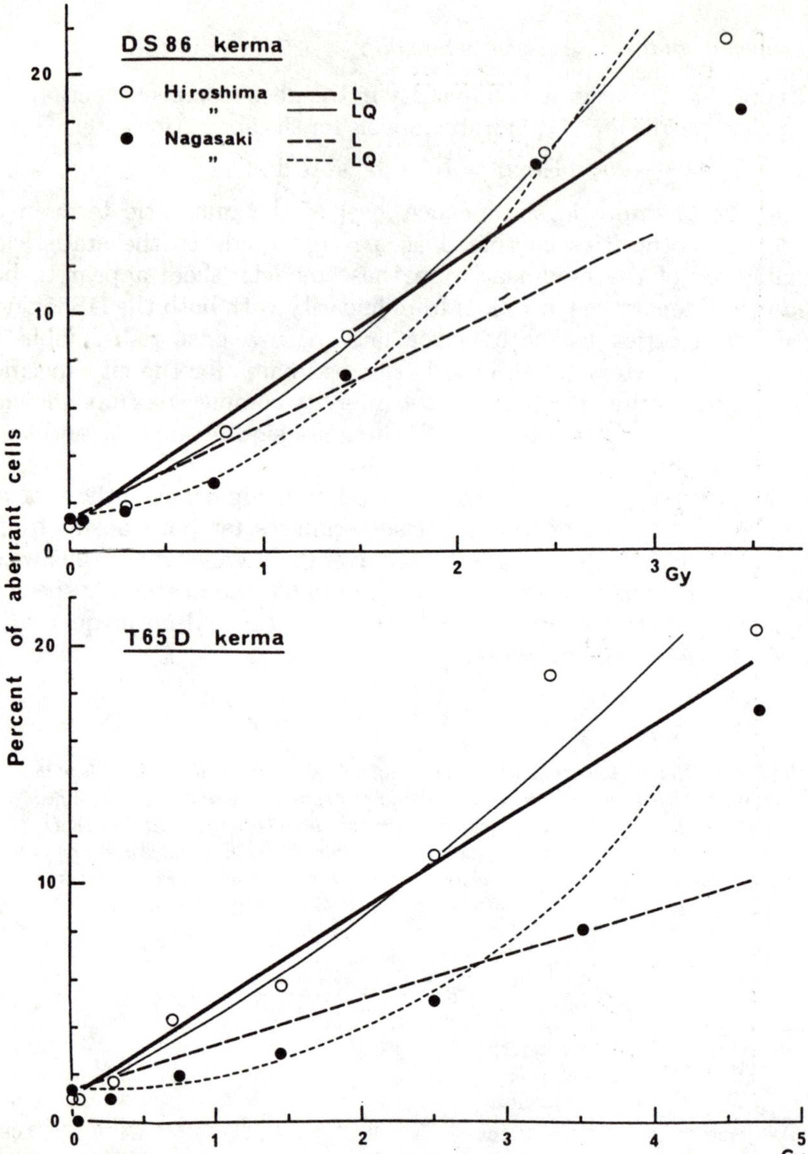

Fig. 1-a

Explanation of Figure

Figure 1 a and b. Graphs, composited from Tables 4, 5 and 8, showing the frequency of aberrant cells in percent plotted against free-in-air tissue kerma (Fig. 1-a), and bone marrow dose estimates (Fig. 1-b) in gray. DS86 (top) and T65D (bottom). Curves are obtained from linear (L) and linear-quadratic (LQ) models. For details, see in the text.

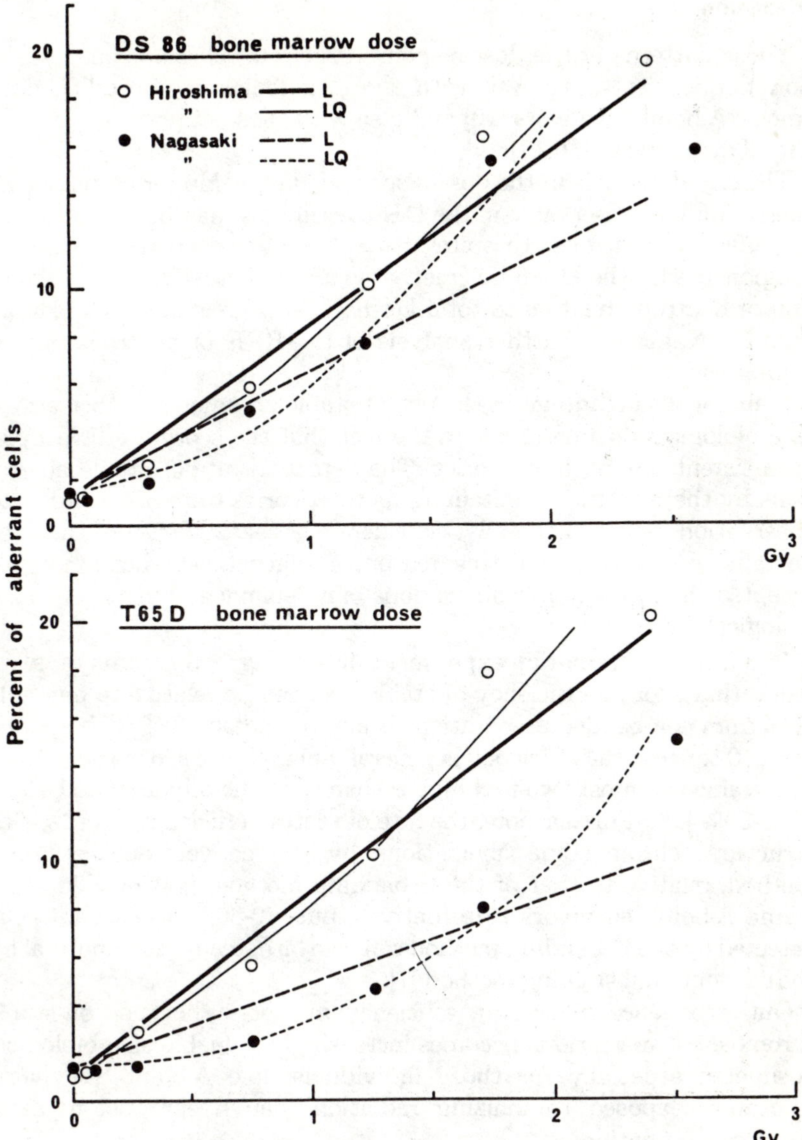
DS 86 bone marrow dose
Hiroshima L
" LQ
Nagasaki L
" LQ
20
10
0
0
1
2
3
Gy
Percent of aberrant cells
T65 D bone marrow dose
20
10
0
0
1
2
3
Gy

Fig. 1-b

Discussion

These patterns of the dose-response curves for chromosome aberration frequencies agree well with those of leukemia mortality data among A-bomb survivors [10] and also with those of previous preliminary data analyses [16,17].

The city difference in the dose-response relationship for chromosome aberration still observable in the DS86 estimates may be attributed to the effect of neutrons. In spite of a substantial decrease of neutron components by the DS86 dosimetry system in Hiroshima, the proportion of neutrons relative to total kerma is still greater for Hiroshima than for Nagasaki. Further analysis of the RBE of neutrons is now in progress.

A major disadvantage in the use of stable chromosome aberrations as a biological dosimeter lies in the fact that the scoring efficiency is inconsistent among laboratories. The detection rate of stable aberrations by the conventional staining method varies from observation to observation but is generally considered to be 20-30% of that for unstable ones [1,18]. For this reason, cytogeneticists have not yet accepted the use of stable aberrations in non-banded preparations as a biological dosimeter.

In our in vitro radiation experiment data using ^{60}Co gamma rays, we found that scoring efficiency of stable aberrations relative to unstable ones from non-banded preparations is approximately 60% with a range from 40 to 70%; the efficiency is generally higher as the dose decreases. This value is almost two-fold higher than the generally accepted value of 20-30% [19]. Furthermore, the rate of scoring efficiency of cells with structural chromosome aberrations by the conventional staining method, relative to that of the G-banding method, is 0.7-0.8 in Hiroshima A-bomb survivors. Alternatively put, 70-80% of aberrant cells detected by the G-banding method can also be detected as abnormal by the conventional staining method [7].

Our experience in scoring efficiency indicates that use of stable chromosome aberrations becomes increasingly valuable as a biological dosimeter, especially for those individuals, like A-bomb survivors, who were exposed to ionizing radiation many years before cytogenetic examination.

References

1. Report of the United Nations Scientific Committee on the Effects of Atomic Radiation. Annex C, Radiation-induced chromosome aberra-

tions in human cells. General Assembly Official Records: Twenty-fourth Session, Suppl. 13(A/7613), United Nations, New York (1969).
2. Awa A.A., Neriishi, S., Honda T., Yoshida M.C., Sofuni T., and Matsui T. Chromosome-aberration frequency in cultured blood-cells in relation to radiation dose of A-bomb survivors. Lancet, ii, 903-905 (1971).
3. Awa A.A., Sofuni T., Honda T., Itoh M., Neriishi S., and Otake M. Relationship between the radiation dose and chromosome aberrations in atomic bomb survivors of Hiroshima and Nagasaki. J. Radiat. Res., 19, 126-140 (1978).
4. Sasaki M.S. Use of lymphocyte chromosome aberrations in biological dosimetry. *In* "Radiation-induced Chromosome Damage in Man," eds. T. Ishihara, M.S. Sasaki, pp. 595-604 (1983). Alan R. Liss, Inc., New York.
5. Sasaki M.S., and Miyata H. Biological dosimetry in atomic bomb survivors. Nature, 220, 1189-1193 (1968).
6. Bender M.A., and Wong R.M.A. Biological indicators of radiation quality. *In* "Reevaluation of Dosimetric Factors: Hiroshima and Nagasaki" eds. V.P. Bond, J.W. Thiessen, pp. 223-240 (1982). Technical Information Center, U.S. Department of Energy.
7. Ohtaki K., Shimba H., Awa A.A., and Sofuni T. Comparison of type and frequency of chromosome aberrations by conventional and G-staining methods in Hiroshima atomic bomb survivors. J. Radiat. Res. 23, 441-449 (1982).
8. Perry P., and Wolff S. New Giemsa method for the differential staining of sister chromatids. Nature 251:156-158 (1974).
9. ISCN (1985): An International System for Human Cytogenetic Nomenclature, eds, D.G. Harnden and H.P. Klinger; Birth Defects: Original Article Series, Vol 21, No 1 (March of Dimes Birth Defects Foundation, New York 1985).
10. Preston D.L., and Pierce D.A. The effect of changes in dosimetry on cancer mortality risk estimates in the atomic bomb survivors, RERF Tech Rep 9-87 (1987).
11. Kerr G.D. Organ dose estimates for Japanese atomic bomb survivors, "Health Physics" 37, pp. 487-508 (1979).
12. Otake M. The nonlinear relationship of radiation dose to chromosome aberrations among atomic bomb survivors, Hiroshima and Nagasaki: RERF Technical Report 19-78 (1978).
13. Otake M. and Prentice R.L. Chromosome aberration analysis based on a Beta-Binomial distribution. RERF Technical Report 4-83 (1983).
14. Moore D.F. Methods of Moments Estimation for overdispersed counts and proportions, Ph.D. dissertation, University of Washington, (1985).
15. McCullagh, P. and Nelder J.A. "Generalized Linear Models" (1983) Chapman and Hall.
16. Awa A.A., Sofuni T., Honda T., Hamilton H.B., and Fujita S. Preliminary reanalysis of radiation-induced chromosome aberrations in relation to past and newly revised dose estimates for Hiroshima and Nagasaki A-bomb survivors. *In* "Biological Dosimetry," eds. W.G. Eisert, M.L. Mendelsohn, pp. 77-82 (1984). Springer-Verlag, Berlin.

17. Fujita S., Awa A.A., Peirce D.A., Kato H., and Shimizu Y. Re-evaluation of biological effects of atomic bomb radiation by changes of estimated dose. Proc. IAEA Symp. "Biological Effects of Low-level Radiation," pp. 55-60 (1983).
18. Buckton K.E., Hamilton G.E., Paton L., and Langland A.O. Chromosome aberrations in irradiated ankylosing spondylitis patients. *In* "Mutagen-induced Chromosome Damage in Man," eds. H.J. Evans and D.C. Lloyd, pp. 142-150 (1978). Edinburgh University Press.
19. Sofuni T., Murata S., Abe T., and Awa A.A. Dose-response relationship of symmetric chromosome exchanges in human lymphocytes exposed to gamma-rays in vitro. "Jap. J. Human Genet." 25:110-111 (1980).

Effects of Ionizing Radiation on The Developing Human Brain

Masanori Otake, Hiroshi Yoshimaru and William J. Schull

Summary

The highest risk of radiation damage to the embryonic and fetal brain occurs 8-15 weeks after fertilization under both the T65DR and DS86 systems. Damage to the 8-15 week old fetus expressed as severe mental retardation appears to be linearly related to the fetal absorbed dose, but to be linear-quadratically related to the fetal absorbed dose for the intelligence score data. Damage to the fetus 16-25 weeks after fertilization suggests a threshold in the low dose region, especially in the DS86 sample.

Introduction

The purposes of this report are twofold, first, to evaluate the absolute risks to the developing human embryonic and fetal brain of exposure to ionizing radiation using the newer DS86 estimates, and, second, to compare the latter estimates with those derived from the earlier T65 dosimetry. Attention will be limited to the data on clinically recognized severe mental retardation, and intelligence as measured by intelligence test scores.

Materials and Methods

Over the years, the Atomic Bomb Casualty Commission and its successor, the Radiation Effects Research Foundation, have established at least three overlapping samples of individuals exposed *in utero* to the atomic bombing of Hiroshima and Nagasaki [1]. These have been commonly termed the original PE-86 sample, the revised or clinical PE-86 sample, and the in utero mortality cohort. Differences between these samples reflect the different purposes for which they were initially chosen, e.g., as the bases for clinical examinations, or mortality surveillance. We present the results of analyses based on two of these, namely, the clinical sample (severe mental retardation) and the original PE-86 sample (IQ) from which the clinical sample was largely, albeit not exclusively derived. The relationship of these samples one to another has been described thoroughly elsewhere [2].

We turn now to a brief description of the data.

Severe mental retardation:

Of the 1,613 nonexposed and exposed children in the in utero clinical sample in Hiroshima and Nagasaki reported by Wood *et al.* [3], 10 cases with unknown dose and 5 cases outside the date of birth restriction have been excluded; thus the sample we use is based on 1,598 individuals [4]. All of the 30 cases of severe mental retardation were diagnosed before the age of 17; these diagnoses were based upon clinical impressions and not on IQ scores, if such existed. A child was judged to be severely mentally retarded if he or she was "unable to perform simple calculations, to make simple conversation, to care for himself or herself, or if he or she was completely unmanageable or had been institutionalized."

Intelligence test (IQ) score:

The intelligence test data involve 1,673 children, excluding those with unknown dose (9 cases) under the T65 system of dosimetry or with no intelligence test results (86 cases) either because of refusal to undergo the test, illness, migration or the like. The routine intelligence testing was conducted in 1955–1956 in the clinical facilities of ABCC in Hiroshima and Nagasaki. Although two intelligence tests, the Tanaka-Binet and the Koga, were in vogue at the time when these children were tested, our analysis is restricted to the Koga test, the much more commonly used with these survivors [5]. Among the 1,673 subjects on whom test results exist, 8 were clinically diagnosed as severely mentally retarded, but were viewed as potentially educable; their IQs varied between 56 and 64.

Dosimetry:

For comparative purposes, we present the results of two analyses, one based on the estimates of fetal absorbed dose using the revised T65 dosimetry, and the other, the DS86 organ dose (uterus). The former are the estimates of the T65 maternal tissue kerma taking into account the relocation of the epicenter in Nagasaki [6] and corrected shielding data. The fetal absorbed doses associated with the T65 system are merely the estimates of maternal tissue kerma multiplied by the average correction factors of Kerr [7]. The DS86 estimates are computed individually without the use of explicit, average correction factors and thus allow better for the scattering of radiant energy that occurs within tissues [8]. Actual fetal absorbed doses as such are not yet available. We use the computed uterus dose. Phantom studies have shown that the correspondence between this dose and that to fetal tissues is high in the latter half of pregnancy, but may overestimate the energy absorbed by embryonic or fetal tissues in the first half when more fluid surrounds the embryo or fetus. The correspondence in estimated doses in the two systems is illustrated in Figure 1 and Table 1.

Neither under the new system nor the old have neutrons been a significant contributor to most fetal exposures, and as a consequence, the effect of this source of radiation is a less pervasive problem than in

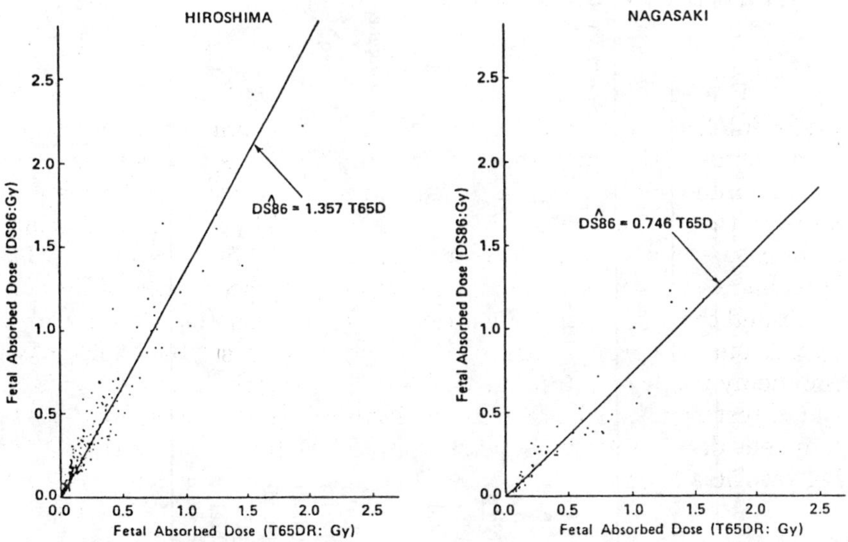

Fig. 1. Relationship between T65 fetal absorbed dose and DS86 uterus organ dose estimates for subsample of severe mental retardation.

TABLE 1—*Relationship of fetal absorbed doses between T65DR and DS86 dosimetry for severe mental retardation and intelligence test scores*

Fetal Absorbed Dose Based on T65DR (Gy)	Fetal Absorbed Dose Based on DS86 (Gy)							Total
	<0.01	0.01-0.09	0.10-0.49	0.50-0.99	1.00+	Subsample	Unknown Dose	
			Severe Mental Retardation					
<0.01	1068	4	1			1073	12	1085
0.01-0.09		191	78			269	23	292
0.10-0.49		6	118	28	1	153	16	169
0.50-0.99			10	12	11	33	1	34
1.00+				2	14	16	2	18
Total	1068	201	207	42	26	1544	54	1598
						(96.6%)	(3.4%)	(100.0%)
			Intelligence Test (IQ) Score					
<0.01	865	47				912	291	1203
0.01-0.09		121	39			160	162	322
0.10-0.49		5	74	16	1	96	15	111
0.50-0.99			9	8	8	25	1	26
1.00+				2	7	9	2	11
Total	865	173	122	26	16	1202	471	1673
						(71.8%)	(28.2%)	(100.0%)

radiation carcinogenesis. Indeed, the free-in-air (FIA) neutron DS86 dose in Hiroshima at about 2,000 m was only 0.04 Gy, and in Nagasaki essentially 0. Thus, the DS86 fetal dose estimates to be discussed shortly include only a rare neutron in Hiroshima, and basically none in Nagasaki and, therefore, the risk estimates depend little upon these particles or their RBE.

Gestational age:

The date of pregnancy ATB is based upon the inferred first day of the last menstrual period, and has been calculated with the following function:

Days of pregnancy (ATB) = 280 - (Date of birth* - 6 or 9 August 1945)

where the mean duration of pregnancy is taken to be 280 days, and * denotes the date of birth obtained by interview with the individual or his or her mother. To obtain the age after fertilization, 14 days have been subtracted from the "days of pregnancy ATB." Age in days was changed to age in weeks by dividing by seven and the latter quotient was presumed to be zero if it was negative.

Statistical Methods

Gestational age is the most important single factor in determining the nature of the insult to the developing embryo or fetus resulting from exposure to ionizing radiation. The analysis of these data must be rooted in this understanding, and not merely in statistical considerations alone. Since different functions in the human brain are localized into different structures, and since the differentiation of these takes place at different stages of gestation and over different periods of time, gestational ages have been grouped so as to reflect these known phases in normal development. Four categories measured from the presumed moment of fertilization have been used: 0–7, 8–15, 16–25; and 26 or more weeks. In the first period, the precursors of the neurons and neuroglia, the two principal types of cells that give rise to the cerebrum have emerged and are mitotically active [9]. In the second, a rapid increase in the number of neurons occurs; they migrate to their ultimate developmental sites and lose their capacity to divide, becoming perennial cells. [10,11] In the third period, differentiation in situ accelerates, synaptogenesis that began about the eighth week increases, and the definitive cytoarchitecture of the brain unfolds. The fourth period is largely one of continued architectural and cellular differentiation and synaptogenesis. This classification has shortcomings, of course; it may confound at least two sources of error—that associated with the presumed moment of fertilization, and that which stems from differences between individuals in their relative stages of development at a fixed interval of time after fertilization. To these uncertainties may be ascribed much of the variation that is seen in response to a given dose.

Given the number of different developmental processes which are proceeding simultaneously and whose relative susceptibilities to radiation-related effects are unknown, a variety of models have been fitted to the individual as well as the grouped data. Depending upon the data set, these include binomial ones, exponential types, and standard regression. Our aims are twofold, namely, to ascertain the most suitable statistical descriptor of the dose-response relationship, and if possible, to contribute to a deeper understanding of the vulnerability of the individual processes. Ultimately, the applicability of any one of these models can be evaluated only by an appeal to experimental findings, to models fitted to other radiation-related biological events, to statistical judgments of "goodness of fit," or to their apparent "reasonableness."

Results

The DS86 sample used in the analysis of severe mental retardation consists of 1,544 (96.6%) of the 1,598 individuals belonging to the clinical sample including all of the 30 diagnosed to be severely mentally retarded. The principal differences between the T65DR and DS86 samples are (1) the shift of 78 (29.0%) of 269 children in the 0.01–0.09 Gy T65 group to the 0.10–0.49 Gy DS86 group, and (2) an increase in the 1.00 + Gy group from 16 in the T65DR to 26 in the DS86.

Insofar as the IQ scores are concerned, the DS86 sample consists of 1,202 (71.8%) of the 1,673 children previously studied; the majority (453) of the 471 individuals on whom estimates are not as yet available had T65 doses of less than 0.10 Gy (291 less than 0.01 Gy), and would presumably have had even smaller DS86 doses. The 1.00 + Gy group increases in number from 9 in the T65DR to 16 in the DS86 estimates. Three of the 24 cases with an IQ score of less than 70 were excluded from the DS86 sample.

Severe mental retardation:

Figure 2 gives the distribution of severely mentally retarded children in the T65DR and DS86 dosimetries by fetal absorbed dose and gestational age in weeks. Note that the average fetal absorbed dose in the successive dose groups is 0, 0.04, 0.23, 0.72 and 1.46 gray for all cases based on the T65DR, and 0, 0.05, 0.23, 0.64 and 1.38 gray for the sample based on the DS86. Observe too that the frequency of severely mentally retarded individuals in the 1.00+ Gy group in the DS86 sample is 12/26 (46.2%) which is higher than the 7/18 (38.9%) in the same T65 dose group for all gestational ages, but the frequency in the 0.50–0.99 Gy DS86 dose group decreases from the 6/34 (17.6%) in the T65DR to 4/42 (9.5%). The number at the top of each histogram is the number of cases with severe mental retardation.

The regression coefficients based on a linear dose response relationship for the T65DR and DS86 samples are summarized by gestational age in Table 2. The maximum likelihood estimates (MLE) of the regression coefficients were obtained by a Newton-Raphson iterative procedure. As is evident from the estimates, the risk of severe mental retardation changes little from one dosimetric system to the other. The highest risk of severe mental retardation due to radiation exposure occurs between 8 and 15 weeks after fertilization, and within this critical period, the occurrence of severe mental retardation appears to be linearly related to the absorbed dose received by the fetus. Overall, the risk per unit dose is four or more times greater during these weeks than in 16–25 weeks. Inspection of the DS86 data also suggest the existence of a threshold in the low dose region for damage in the period 16–25 weeks.

It has been suggested [12] that the model proposed by Lea [13], based upon cell survival considerations might be more suitable than a simple linear model. Accordingly, we have fitted an exponent model to these same data (see Table 3). The results are essentially the same as those in the linear model, but the absolute risk based on an exponent model is slightly greater than that of the linear model.

Intelligence test scores:

Figure 3 presents graphically the specific frequency distributions of the intelligence test scores by gestational age and fetal absorbed dose group. When the homogeneity of the variances was examined among exposure categories for fixed gestational age groups, Bartlett's chi-squares (each with 4 degrees of freedom) ranged between 2.65 and 5.49

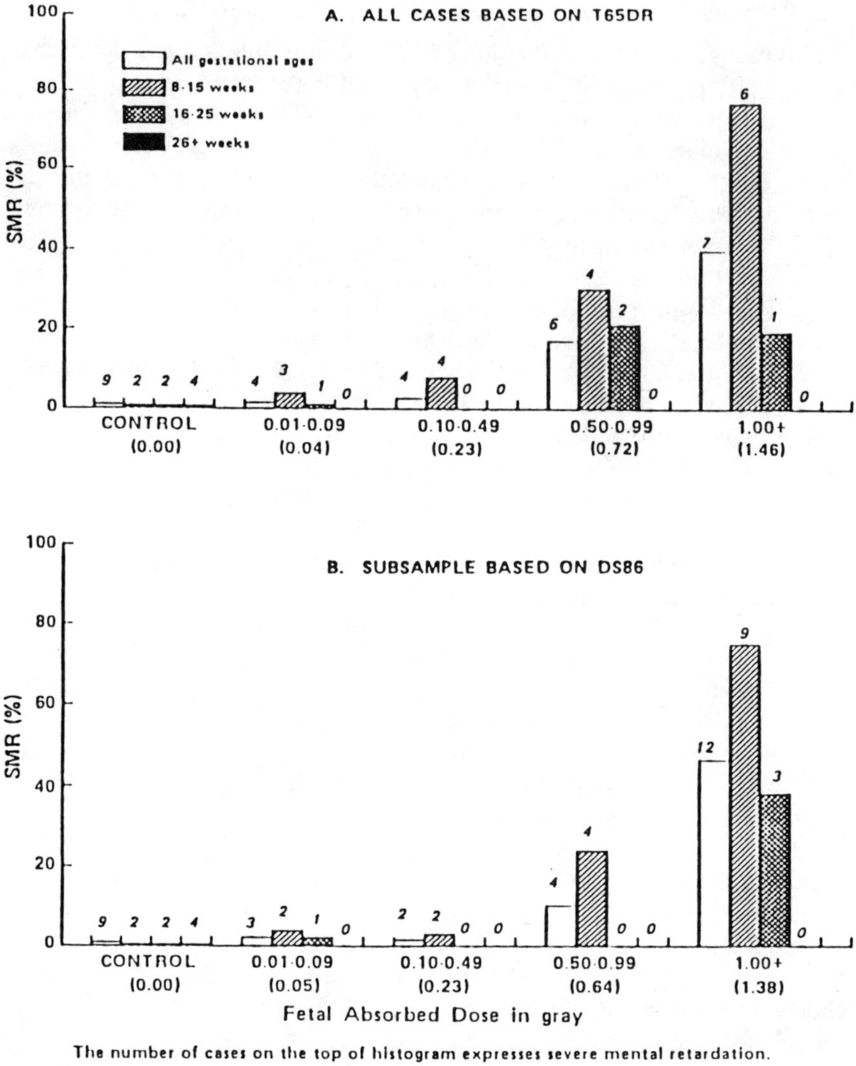

Fig. 2. Severe mental retardation frequency and absorbed dose with both cities combined

for the DS86 sample with or without inclusion of the eight severely retarded individuals and between 2.10 and 14.02 for all cases in the T65DR sample. Of these, only one age-group, 16–25 weeks based on

TABLE 2—*Relationship of severe mental retardation to fetal absorbed dose, T65DR and DS86 estimates*

Gestational Age	MLE of Regression Coefficients				Goodness of Fit	
	a (10^{-2})	Sa (10^{-2})	b (Gy)	Sb (Gy)	χ^2	P
			A. All cases based on T65DR			
All gestational ages	0.762	0.241	0.183**	0.043	4.08	0.25
8-15 weeks	0.866	0.571	0.462**	0.091	1.55	0.67
16-25 weeks	0.603	0.395	0.101 Sugg	0.058	2.04	0.56
			B. Subsamples based on DS86			
All gestational ages	0.736	0.237	0.166**	0.038	13.04	0.01
8-15 weeks	0.747	0.511	0.429**	0.088	4.78	0.19
16-25 weeks	0.608	0.402	0.095 Sugg	0.054	6.90	0.08

Linear model: $P_i = a + bD_i (i = 1,2,...,5)$

**Significance at < 0.01 level, * at < 0.05 level and sugg at < 0.10 level.

TABLE 3—*Relationship of severe mental retardation to fetal absorbed dose, T65DR and DS86 estimates*

Gestational Age	MLE of Regression Coefficients				Goodness of Fit	
	a (10^{-2})	Sa (10^{-2})	b (Gy)	Sb (Gy)	χ^2	P
			A. All cases based on T65DR			
All gestational ages	0.763	0.243	0.194**	0.049	4.97	0.17
8-15 weeks	0.824	0.559	0.560**	0.149	2.66	0.45
16-25 weeks	0.604	0.398	0.104 Sugg	0.063	2.10	0.55
			B. Subsamples based on DS86			
All gestational ages	0.742	0.240	0.169**	0.041	15.78	0.01
8-15 weeks	0.746	0.516	0.479**	0.124	7.09	0.07
16-25 weeks	0.612	0.406	0.095 Sugg	0.056	7.35	0.06

Exponent linear model: $P_i = 1 - \exp - (a + bD_i)$

the T65DR was significant at the 1% level. When we combined the 0.10–0.49 Gy and 0.50–0.99 Gy groups because of their small numbers, Bartlett's test for the 16–25 weeks was not significant ($x2 = 1.19$, $df = 3$). There is, therefore, no compelling evidence that the variances are different either among doses for fixed gestational age, or among gestational ages for fixed doses.

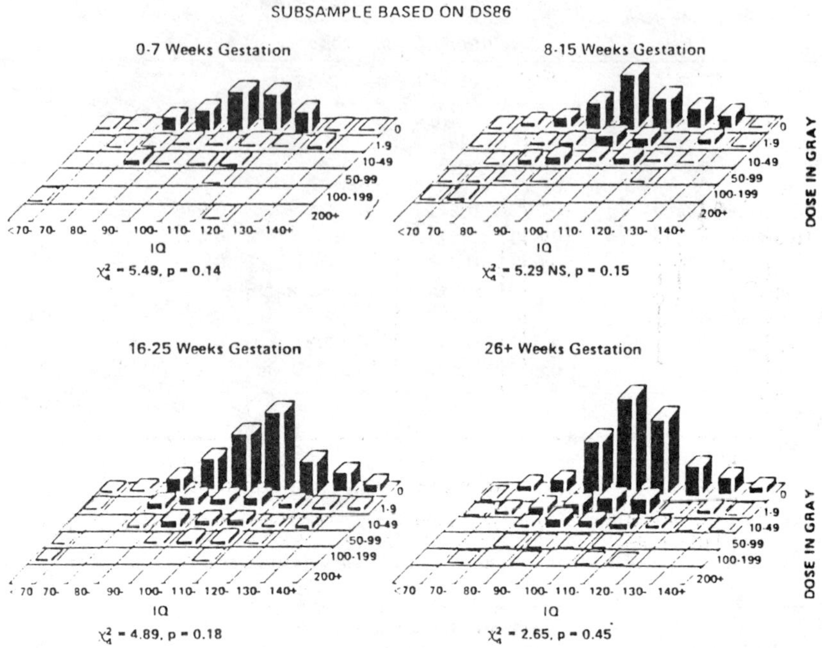

Fig. 3. The distribution of intelligence test scores by gestational age ATB and estimated fetal absorbed dose

Normality of the distribution of intelligence test scores was examined using the skewness test proposed by Fisher. The 20 measures of skewness in the age-exposure cells, ranged from -0.68 to 0.45 for the T65DR and -0.96 to 0.34 for the DS86 sample, and none deviated significantly from zero. These findings suggest that that the data are drawn from a single unimodal probability distribution; alternatively, there is no evidence of a commingling of two or more distributions which could be interpreted as supporting the notion of groups of individuals with differing sensitivities to ionizing radiation. Figure 4 presents the mean intelligence scores with their confidence intervals for the various gestational age groups. Note the systematic decline in the 8–15 week group, and to lesser extent the 16–25 week group.

A better analytic approach, one that does not depend upon grouping the data, is to fit a regression of intelligence test scores on the individual estimates of exposure within an age group. The regression coefficients obtained when a linear dose-response model is fitted are presented in Table 4. Significant linear coefficients occur for the 8–15

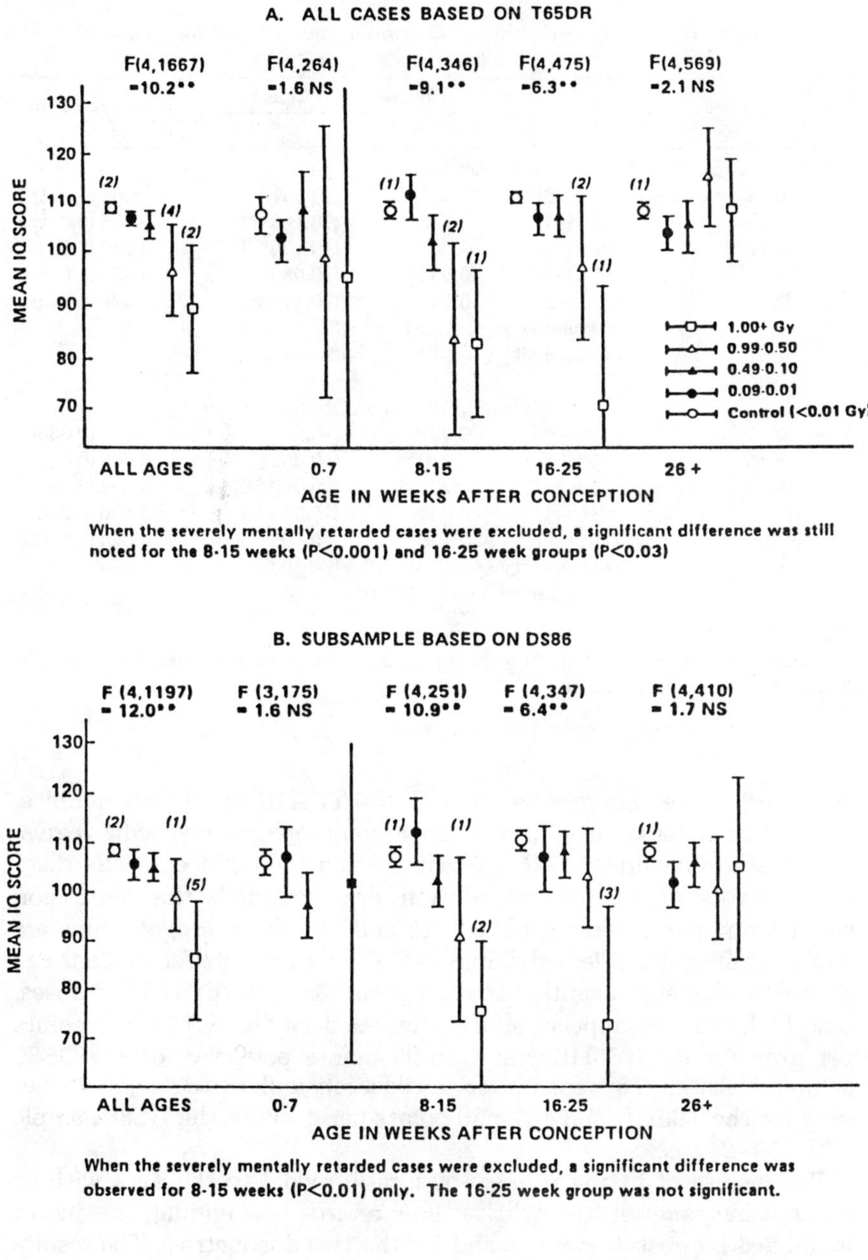

Fig. 4. Mean IQ score and 95% confidence limits by gestational age in weeks and fetal absorbed dose

TABLE 4—*Regression coefficients of IQ score on fetal absorbed dose based on a linear model*

Age Group in Weeks	n	Regression Coefficient b		Standard Error Sb
		A. All cases based on T65D		
0- 7	269 (0)	-0.057	(-0.057)	0.061 (0.061)
8-15	350 (-4)	-0.212**	(-0.168**)	0.036 (0.042)
16-25	480 (-3)	-0.207**	(-0.133*)	0.042 (0.053)
26 +	574 (-1)	-0.033	(-0.035)	0.055 (0.054)
Total	1673 (-8)	-0.155	(-0.111**)	0.023 (0.025)
	Heterogeneity $\chi^2 = 11.07$ df = 3 p = 0.01			
	(Heterogeneity $\chi^2 = 4.75$ df = 3 p = 0.19)			
		B. Subsample based on DS86		
0- 7	179 (-0)	-0.027	(-0.27)	0.053 (0.053)
8-15	256 (-4)	-0.290**	(-0.250**)	0.042 (0.051)
16-25	352 (-3)	-0.204**	(-0.098 Sugg)	0.044 (0.057)
26 +	415 (-1)	-0.042	(-0.044)	0.050 (0.050)
Total	1202 (-8)	-0.158**	(-0.102**)	0.024 (0.026)
	Heterogeneity $\chi^2 = 22.30$ df = 3 p = 0.01			
	(Heterogeneity $\chi^2 = 11.82$ df = 3 p = 0.0)			

The coefficients are expressed as change in IQ points per 0.01 Gy.
Numbers and estimates in parentheses reveal the results after exclusion of clinically diagnosed cases of retardation.

and 16–25 week age groups for both the T65DR and DS86 samples. Significant heterogeneity does exist among age-group-specific regression coefficients under both systems. When the eight clinically diagnosed cases of mental retardation are excluded, the regression coefficients associated with the 8–15 and 16–25 week age groups are again significant but less strikingly so. The coefficients for the four age groups are not significantly heterogeneous based on the T65 DR doses, but the loss in IQ appears substantial—at face value, 16 to 21 points per gray for the T65DR and 8 to 29 points per gray for the DS86 sample in the 8–15 weeks after fertilization and 13 to 21 points per gray for the T65DR and 10 to 20 points per gray for the DS86 sample in the 16–25 weeks.

The results of fitting a linear-quadratic model to the data with or without inclusion of the eight severely retarded individuals are shown in Table 5 by gestational age and for the two dosimetries. The results based upon the T65DR show that the only regression coefficients significantly different from zero are for the 8–15 week age group; this holds true whether the mentally retarded are or are not included, but

TABLE 5—*Regression coefficients of IQ score on fetal absorbed dose based on a linear-quadratic model*

Age Group in Weeks	n	Linear Coefficient b (0.01 Gy)	Standard Error Sb (0.01 Gy)	Quadratic Coefficient c (Gy)	Standard Error Sc (Gy)
		A. All cases based on T65DR			
0- 7	269 (0)	-0.101 (-0.101)	(0.166)	0.038 (0.038)	0.134 (0.134)
8-15	350 (-4)	-0.397** (-0.287**)	0.093 (0.100)	0.112* (0.068)	0.051 (0.052)
16-25	480 (-3)	-0.137 (-0.065)	0.087 (0.124)	-0.061 (-0.087)	0.066 (0.145)
26 +	574 (-1)	-0.150 (-0.156)	0.127 (0.126)	0.133 (0.137)	0.130 (0.130)
Total	1673 (-8)	-0.169** (-0.119*)	0.049 (0.050)	0.011 (.006)	0.033 (0.035)
	Heterogeneity	$\chi^2 = 5.38$ df = 3	p = 0.15	$\chi^2 = 4.70$ df = 3	p = 0.20
	(Heterogeneity	$\chi^2 = 2.24$ df = 3	p = 0.15	$\chi^2 = 1.44$ df = 3	p = 0.70)
		B. Subsample based on DS86			
0- 7	179 (0)	-0.340* (-0.340*)	0.135 (0.135)	0.164* (0.164*)	0.065 (0.065)
8-15	256 (-4)	-0.258* (-0.261*)	0.111 (0.122)	-0.029 (0.012)	0.092 (0.117)
16-25	352 (-3)	-0.036 (-0.059)	0.093 (0.156)	-0.168* (-0.062)	0.082 (0.233)
26 +	415 (-1)	-0.101 (-0.108)	0.128 (0.127)	0.071 (0.077)	0.142 (0.140)
Total	1202 (-8)	-0.207** (-0.190**)	0.048 (0.048)	0.042 (0.081*)	0.036 (0.038)
	Heterogeneity	$\chi^2 = 4.57$ df = 3	p = 0.20	$\chi^2 = 10.52$ df = 3	p = 0.01
	(Heterogeneity	$\chi^2 = 2.67$ df = 3	p = 0.45	$\chi^2 = 1.99$ df = 3	p = 0.58)

Numbers and estimates in the parenthesis reveal the results after exclusion of clinically diagnosed cases of retardation.

the quadratic coefficients are positive. Homogeneity of the regression coefficients among the four gestational age groups were examined using Armitage's method [13]. When tested separately, both the linear and the quadratic terms show no significant heterogeneity with the T65DR doses. However, significant heterogeneity was observed in the DS86 sample when the severely mentally retarded cases are included. Both linear and quadratic terms in the 0–7 and 8–15 weeks of gestation are significant, but the quadratic terms in the two gestational groups are positive. A significant negative quadratic term was noted only for the group 16–25 weeks after fertilization. After exclusion of the severely retarded cases, no significant difference is seen in the quadratic term and in the heterogeneity of regressions.

Conclusions

(1) The highest risk of radiation damage to the embryonic and fetal brain occurs 8–15 weeks after fertilization in both the T65DR and DS86 samples.

(2) The damage to the 8–15-week-old fetus appears to be linearly related to the fetal absorbed dose for severe mental retardation data, but to be linear-quadratically related to the fetal absorbed dose in the intelligence score data.

References

1. Beebe, G.W. and Usagawa, M. (1968). "The Major ABCC Samples," ABCC Technical Report No. 12-68.
2. Schull, W.J. and Otake, M. (1968). "Effects on Intelligence of Prenatal Exposure to Ionizing Radiation," RERF Technical Report No. 7-86.
3. Wood, J.W., Johnson, K.G., Omori, Y., Kawamoto, S., and Keehn, R.J. (1967). "Mental Retardation in Children Exposed In Utero, Hiroshima and Nagasaki," Am. J. Public Health 57, 1381–90. (ABCC Technical Report No. 10-66).
4. Otake, M., Yoshimaru, H. and Schull, W.J. (1987). "Severe Mental Retardation among the Prenatally Exposed Survivors of Atomic Bombing of Hiroshima and Nagasaki: A Comparison of the T65DR and DS86 Dosimetry Systems," (RERF Technical Report No. 16–87).

5. Schull, W.J., Otake, M. and Yoshimaru, H. (1988). "Effect on Intelligence Test Score of Prenatal Exposure to Ionizing Radiation in Hiroshima and Nagasaki: A Comparison of the T65DR and DS86 Dosimetry Systems," (RERF Technical Report No. 3-88).
6. Kerr, G.D. and Solomon, D.K. (1976). "The Epicenter of Nagasaki Weapon: A Reanalysis of Available Data with Recommended Values," ORNL-TM-5319, Oak Ridge National Laboratory.
7. Kerr, G.D. (1979). "Organ Dose Estimates for the Japanese Atomic Bomb Survivors," Health Phys. 37, 487–508.
8. Woolson, W.A., Egbert, S.D. and Gritzner, M.L. (1986). "Report 10 Dosimetry System, 1986 (DS86) for Japanese A-bomb Survivors," DRAFT SAIC-86/1027.
9. Martinez Martinz, P.F.A. (1982). "Neuroanatomy: Development and Structure of the Central Nervous System," Philadelphia, WB Saunders Co.
10. Rakic, P. (1975). "Cell Migration and Neuronal Ectopias in the Brain," In Morphogenesis and Malformation of the Face and Brain. Ed. by Bergsma D. Alan Liss, New York, 95–129.
11. Rakic, P. (1978). "Neuronal Migration and Contact Guidance in the Primate Telencephalon," Postgrad. Med. J. 54 (Supple) 1, 25–40.
12. Mole, R.H. (1987). "Irradiation of the embryo and fetus," Brit. J. Radiol. 60, 17–31.
13. Lea, D.E. (1955). "Actions of Radiations on Living Cells," 2nd Edition, New York: Cambridge University Press.
14. Armitage, P. (1971). "Statistical Methods in Medical Research," 3rd Edition, 279–81. John Wiley & Sons, New York.

Implications of the New Dosimetry for Risk Estimates

William J. Schull, Hiroo Kato, Yukiko Shimizu

Introduction

Over the years, the Atomic Bomb Casualty Commission, and its successor, the Radiation Effects Research Foundation, have used a variety of measures of individual exposures to the atomic bombing of Hiroshima and Nagasaki; these have included distance from the hypocenter (Moloney and Kastenbaum, 1954), the presence or absence of the symptoms associated with acute radiation illness (Neel and Schull, 1956), and now no less than three separate physical dosimetries, namely the T57 (Ritchie and Hurst, 1959; Arakawa, 1960; for an application of these doses, see Ishida and Jablon, 1964), the T65 (Milton and Shohoji, 1968), and the DS86 (Roesch, 1987). The introduction of each successive method of expressing doses has of necessity led to the reestimation of the risk coefficients associated with the various known effects of ionizing radiation, and a reexamination of the dose-response relationships. Moreover, with these changes there has emerged an ability to estimate the risk coefficients in terms other than kerma, whether free-in-air or in house (shielded). The purpose of this conference has been to examine the changes in these estimates, whether expressed as absolute or relative risk coefficients, occasioned by the introduction of the newly installed system of dosimetry, the so-called DS86, and the aim of this specific presentation is to summarize these changes and briefly address their implications.

Changes in dose

It is perhaps well to begin this task by reiterating the principal differences that obtain between the T65 and the DS86 dosimetries. Briefly, they are as follows:

1) The consensus yield of the Hiroshima weapon is now presumed to have been approximately twenty percent greater than earlier thought, that is, 15 rather than 12.5 kilotons.

2) Although the free-in-air (FIA) gamma doses are somewhat greater at distances of 1.4 kilometers or more, neutron exposures are less in both cities, and substantially so in Hiroshima. Indeed, at about 10% of their previously estimated value in the latter city, they are now so small as to make direct estimates of the neutron RBE extremely difficult, if not impossible.

3) Attenuation of the FIA gamma kerma by wooden Japanese structures, houses and tenements, is approximately two-fold greater under the DS86 than under the T65 dosimetry; however, attenuation of the neutron kerma by such structures differs little.

4) Transmission through tissue of gamma rays is significantly higher, at least for the deeply situated organs than heretofore supposed (Kerr, 1979).

5) Finally, it warrants note that for some 18% or so of the members of the Life Span Study cohort, largely individuals exposed in concrete buildings or factories, doses cannot as yet be computed, and the new dosimetry does not clarify all of the implausibly high exposures seen with the T65 system. There remains a substantial number of individuals whose estimated whole-body shielded kerma exposures exceed 4 Gy, many greater than 6 Gy. These are doses at or above the recently estimated LD_{95} in these cities (Fujita, Kato and Schull, 1987), and given the virtual obliteration of the immune system at doses in excess of 7 or 8 Gy, make survival under the circumstances that obtained in these cities most unlikely. Moreover, as Dr. Awa has noted (this conference), for some of these individuals at least the physical estimate is at sharp variance with the chromosomal findings or the data on the frequency of somatic mutations or both. There is obviously a need to develop better means to address these incongruities than the simple truncation of dose since their inclusion in analyses can affect the shape of the dose-response relationship as well as estimates of its parameters (see Jablon, 1971; Gilbert, 1984; Gilbert and Ohara, 1984).

Changes in risk

To turn now to estimates of risk of malignancy, as you will have observed from the presentation of Dr. Preston, these changes have resulted in excess relative risks for leukemia, and all cancers except leukemia, when averaged over six categories of age at the time of bombing (ATB) and sex, that are about 75–85% higher than the comparable T65 estimates if dose is expressed in terms of total shielded kerma (Preston and Pierce, 1987) and the comparison is made with the full T65 cohort, some 18% or so of whom have no DS86 doses as previously noted. However, when expressed in terms of organ dose, the differences between the two dosimetries are substantially smaller. Parenthetically, it should be noted that with the DS86 system the anomalous absence of leukemia cases in Nagasaki in the 0.5–1.0 Gy dose range no longer obtains. The excess relative risk per ten thousand person-year-gray over the entire follow-up period for leukemia is 3.52 and 3.46 for the T65 and DS86 dosimetries, respectively, if the data are constrained to those individuals with estimated exposures of less than 4 Gy and marrow doses are used. The comparable values for all cancers except leukemia, using intestinal dose, are 15.81 and 13.03, respectively. Under both dosimetries estimates of the effects of gamma dose decrease with increasing assumed values of the relative biological effectiveness of neutrons, but less pronouncedly so with the new dosimetry.

Preliminary data are available on the relative risk coefficients, the excess deaths per ten thousand person-year-gray, and attributable risk associated with specific sites of malignancy (see Table 1). These data suggest that in terms of total shielded kerma the dose-response curve is more linear for virtually all sites, and the difference in cancer mortality between Hiroshima and Nagasaki is smaller than earlier reported, and no longer statistically significant. The excess relative risks are 1.35–1.51 fold higher and the excess numbers of cancer deaths are increased 1.43–1.49 fold. Specifically, for both cities and sexes, and all ages ATB, the excess relative risk at one gray (defined as the relative risk coefficient minus one) are as follows (see Table 1 for their confidence intervals): leukemia (3.97), esophagus (0.43), stomach (0.23), large intestine except rectum (0.56), lung (0.46), breast (1.02), ovary (0.80), bladder (1.06), and multiple myeloma (1.89). The attributable risks do not differ, however, between the two dosimetric systems; the ratio of the DS86 and T65 estimates varies from 0.95 to 1.04. When couched in total organ dose (see Table 2), the excess relative risk at one gray for cancers of the stomach, colon, lung and breast are 0.27, 0.85, 1.64, and 1.12, respectively. The ratio of these values to the corresponding T65 ones range between 0.86 and 1.15. The excess number of cancer deaths per

TABLE 1—*Current estimates of the risk of site-specific malignancies based upon mortality in the Life Span Study sample in the years 1950 to 1985. These estimates are based upon shielded kerma, a linear response model over the whole dose range, and only those members of the sample on whom DS86 doses exist (both cities, sexes, and all ages ATB combined). This table has been adapted from Table 5 in Shimizu, et al., 1987.*

Site of cancer	Excess relative risk per gray	Excess deaths per 10^4 PYGy	Attributable risk (%)
Leukemia	3.97 (2.89, 5.39)	2.30 (1.88, 2.73)	56.6 (46.3, 67.1)
All except leukemia	0.30 (0.23, 0.37)	7.49 (5.90, 9.15)	8.0 (6.3, 9.8)
Esophagus	0.43 (0.09, 0.92)	0.34 (0.08, 0.67)	12.8 (3.0, 25.0)
Stomach	0.23 (0.13, 0.34)	2.09 (1.20, 3.06)	6.4 (3.6, 9.3)
Colon	0.56 (0.25, 0.99)	0.56 (0.26, 0.91)	15.2 (7.0, 24.9)
Lung	0.46 (0.25, 0.72)	1.26 (0.70, 1.89)	11.6 (6.5, 17.4)
Breast	1.02 (0.48, 1.76)	1.04 (0.53, 1.61)	22.4 (11.5, 35.0)
Ovary	0.80 (0.14, 1.85)	0.45 (0.09, 0.89)	18.6 (3.6, 37.1)
Bladder	1.06 (0.46, 1.09)	0.56 (0.27, 0.90)	23.4 (11.2, 37.4)
Multiple myeloma	1.89 (0.56, 4.45)	0.22 (0.08, 0.39)	32.9 (11.5, 59.8)

( ): 90% confidence interval

TABLE 2—*Current estimates of the risk of site-specific malignancies based upon mortality in the Life Span Study sample in the years 1950 to 1985. These estimates are based upon organ doses, a linear response model over the whole dose range, and only those members of the sample on whom DS86 doses exist (both cities, sexes, and all ages ATB combined). This table has been adapted from Table 6 in Shimizu et al., 1987.*

Site of cancer	Excess relative risk per gray	Excess deaths per 10^4 PYGy	Attributable risk (%)
Leukemia	5.21 (3.84, 7.14)	2.95 (2.44, 3.50)	58.5 (48.5, 69.5)
All except leukemia	0.41 (0.32, 0.51)	10.18 (7.98, 12.46)	8.2 (6.4, 10.0)
Esophagus	0.06 (0.01, 0.12)	0.45 (0.11, 0.88)	13.0 (3.0, 25.5)
Stomach	0.27 (0.14, 0.42)	2.42 (1.25, 3.70)	5.7 (2.9, 8.7)
Colon	0.85 (0.30, 1.45)	0.81 (0.40, 1.30)	16.3 (8.0, 26.2)
Lung	0.64 (0.36, 0.98)	1.71 (0.99, 2.51)	12.6 (7.2, 18.4)
Breast	1.12 (0.50, 1.94)	1.14 (0.55, 1.80)	21.5 (10.4, 33.9)
Ovary	0.14 (0.04, 0.30)	0.74 (0.24, 1.36)	22.6 (7.2, 41.9)
Bladder	0.13 (0.05, 0.24)	0.68 (0.31, 1.12)	21.5 (9.8, 35.7)
Multiple myeloma	0.23 (0.07, 0.53)	0.26 (0.09, 0.47)	31.8 (11.0, 57.6)

( ): 90% confidence interval

10^4 PYGy for these same cancers is 2.42, 0.81, 1.71 and 1.14. Again the estimates of attributable risk are little affected.

Although, as previously noted, the neutron doses are now much smaller, analyses using the gamma and neutron doses separately reveal that the dose-response curve can be satisfactorily represented by a model that is linear in both gamma rays and neutrons; however, precise estimation of the effects of neutrons and their RBE is difficult, if not impossible. For fixed RBEs of 1, 10 and 20, the "pure" gamma coefficients at 1 gray are 1.3–2.0 fold greater than their comparable T65 estimates. Estimates of these coefficients using variable RBEs do not differ appreciably.

The average percentage of cells with chromosomal aberrations, when adjusted for extra-binomial variability, increases less rapidly with dose in Nagasaki than in Hiroshima under both dosimetric systems; however, the difference in slopes between the cities is less for the DS86 than the T65 doses although it remains statistically significant for both shielded kerma and bone marrow doses. The DS86 slope for shielded kerma is about 60% larger than the corresponding T65 slope, but for marrow doses there is little change. Further analyses to test nonlinearity in the dose-response function show that there appear to be significant departures from linearity with both dosimetries and for both shielded kerma and bone marrow dose.

Insofar as effects on the developing human embryonic or fetal brain are concerned, we have seen that the change in dosimetry has little effect (see Table 3). The period of maximum vulnerability remains the same under both systems of dosimetry, and the estimates of risk are little changed. Within the most vulnerable period, 8–15 weeks, the risk of severe mental retardation at one gray is approximately 45% under a simple linear model, and about 50% under an exponential-linear one.

Biological understanding of the bases of risk

The risk coefficients to which we have alluded serve at least two non-mutually exclusive purposes, namely, as guidance to regulatory agencies charged with establishing permissible occupational and population exposures, and as a means to the understanding of the biological events, cellular and molecular, which presumably subtend the effects that are observed. Although it would be foolhardy to presume that further dosimetric refinements would not occur, and thus some continued sharpening of the risk coefficients would not obtain, generally speaking even now we can better estimate the probable number of

TABLE 3—*Dose-response relationship of severe mental retardation to "fetal" absorbed doses under two different models based on the DS86 dosimetry*

Gestational age	Cities combined					
	a	b	s_b	X^2_{Reg}	X^2_{Res}	P_{Res}
SIMPLE LINEAR MODEL						
A. Relationship to fetal absorbed dose						
All gestational ages	0.742	0.172	0.039	18.91**	13.94	<0.01
8–15 weeks	0.749	0.443	0.090	24.17**	4.47	0.21
16–25 weeks	0.611	0.097	0.055	3.11	6.55	0.09
B. Relationship to fetal absorbed dose when all "controls" are excluded						
8–15 weeks	0.209	0.456	0.100	20.60**	4.54	0.11
16–25 weeks	0.391	0.102	0.060	2.89	6.56	0.04
C. Relationship to fetal absorbed dose when all "controls" are combined						
8–15 weeks	0.832	0.441	0.090	24.18**	4.47	0.21
EXPONENTIAL LINEAR MODEL						
A. Relationship to fetal absorbed dose						
All gestational ages	0.749	0.175	0.043	16.76**	15.97	<0.01
8–15 weeks	0.747	0.498	0.129	15.00**	6.70	0.08
16–25 weeks	0.616	0.097	0.057	2.90	7.00	0.07
B. Relationship to fetal absorbed dose when all "controls" are excluded						
8–15 weeks	−0.016	0.523	0.150	12.18*	6.70	0.04
16–25 weeks	0.020	0.105	0.065	2.61	8.05	0.02
C. Relationship to fetal absorbed dose when all "controls" are combined						
8–15 weeks	0.835	0.496	0.128	15.01**	6.70	0.08

a is the estimated number of cases of mental retardation (per 100 persons) in the zero-gray group; b is the increase in the frequency of severe mental retardation with dose expressed in gray and s_b its standard error. The chi-squares are those associated with the regression and residual variability, respectively, and P is the probability (two-tailed) of exceeding the residual chi-square by chance under the null hypothesis.

cancer deaths (or other effects) induced by a given dose of ionizing radiation, than we can understand the cellular and molecular processes which radiation initiates or promotes. This situation seems unlikely to change until the analytic models fitted to the observational data better incorporate the biology, admittedly limited, that is known, and a more dynamic interaction occurs between the epidemiologically oriented and molecular and cellular biologists. An allegation of this nature, if it is not to be solely gratuitous, requires some alternatives.

Let me try to illustrate briefly possible avenues through which such an interaction could occur. First, the analyses of the leukemia data, from Japan and elsewhere, have treated this malignancy as if it was homogeneous, and have largely ignored the known differences that exist in the various leukemias. Perhaps, this is no more clearly shown than in the treatment of childhood leukemia as if it differed from adult leukemias only in the age of onset. This is patently not so. Childhood leukemias differ from adult ones in their cell types, in their chromosomal markers, and in their response to therapy. All of these observations suggest a differing set of cellular or molecular events, and thus the possibility of different effects of irradiation. Second, it is increasingly clear that oncogene amplification is associated with the occurrence of many site-specific malignancies. Indeed, so many now that it is difficult to know whether it is causally related or merely a consequence of the transformation of the normal cell to a malignant state. While the demonstration of oncogene amplification is still too tedious to be applicable to large numbers of individuals at risk, this seems unlikely to continue to be so for long. It would, therefore, behoove us to consider how to incorporate newer molecular and cellular information into the risk models we fit. Attempts to achieve this have been few. Third, there remains much to be done to identify modifiers of risk, and to quantitate their effects. For example, the role of individual differences in genetic background in the enhancement or amelioration of risk hangs like an albatross. It is invariably cited as a possible modifier, but little direct evidence is available save for a small group of rare, recessively inherited diseases. As a geneticist, it is difficult to believe that the effects of genetic differences are not more pervasive. Fourth, as the number of cell markers increase through which a malignant clone can be characterized, it becomes increasingly practicable to determine better whether radiation-induced malignancies are indeed the same or different qualitatively or quantitatively from those that arise spontaneously. Some experimental evidence suggests they are indeed different quantitatively.

Obviously numerous other alternatives could be cited, but to make more clear the possible breadth and fruitfulness of the interaction consider the issue of brain damage following prenatal exposure to ionizing radiation. Few areas of research have seen such rapid recent advances in understanding and mensurational techniques as neurobiology yet surprisingly little of this newer information on cortical histogenesis and function serves to guide epidemiological studies. As we have seen (Otake, Yoshimaru and Schull, this Conference), the occurrence of severe mental retardation, with or without accompanying small head size, has been the most conspicuous effect of prenatal exposure to the atomic bombing of Hiroshima and Nagasaki yet we know virtually nothing about the underlying cellular or molecular events which gave rise to these observations. Heterotopic gray matter has been known to be associated with other abnormalities of brain development, structure, and function for over a century, but its prevalence and importance have been difficult to document. Recognition of heterotopic areas rested solely on autopsy findings until the introduction of pneumographic techniques which are not only invasive but unsatisfactory in many respects. Computed tomography improved matters somewhat, but the introduction of magnetic resonance imaging with its greater definition of gray and white matter has precipitated a flurry of interest and case studies, and has opened opportunities to document the occurrence of brain abnormalities previously unavailable. As yet the data associating heterotopic gray matter with the occurrence of severe mental retardation is limited. Cost and time presently preclude the use of brain imaging in most epidemiological settings, but its judicious application could, for example, materially improve understanding of specific brain developmental processes, such as neuronal migration, through the better delineation of the factors which intrude on the normal process.

Similarly it grows steadily more practicable to examine cortical function from a variety of perspectives, and to entertain at least the possibility of demonstrating radiation-related lesions undemonstrable through clinical examinations alone. If small doses at critical stages in development can impair neuronal migration, as much experimental evidence suggests (UNSCEAR, 1986), surely there will exist individuals wherein the damage is not sufficiently great to produce a gross sociological handicap but would be measurable by these newer techniques.

As these examples have attempted to convey, risk estimation should be seen as a dynamic process, one that draws simultaneously on developments in statistics and in biology generally, and seeks to provide insight into the origins of these effects we observe. This suggests, of course, that estimates will change as understanding grows, a situation which may not please the regulator who is apt to see in this dynamism a threat to authority and acceptance of the notion of standards if these undergo alteration frequently. Ostensibly, however, our aims are to prevent as many of the untoward consequences of exposure to ionizing radiation as practicable; prevention rests on understanding not on statistical elegance.

References

1. Arakawa, E.: Radiation dosimetry in Hiroshima and Nagasaki atomic bomb survivors. N Engl J Med 263:488–493, 1960
2. Fujita, S., Kato, H., and Schull, W.J.: The LD_{50} associated with exposure to the atomic bombing of Hiroshima and Nagasaki: A review and reassessment. Radiation Effects Research Foundation Technical Report, 1987
3. Gilbert, E. and Some effects of random dose measurement errors on analysis of atomic bomb survivor data. Radiat Res 98:591–605, 1984
4. Gilbert, E. and Ohara, J.: An analysis of various aspects of atomic bomb dose estimation at RERF using data on acute radiation symptoms. Radiat Res 100:124–139, 1984
5. Ishida, M. and Jablon, S.: Report 4. Mortality in a-bomb survivors by age cohorts, 1950–1959. Atomic Bomb Casualty Commission Technical Report 14–64, 1964
6. Jablon, S.: Atomic bomb radiation dose estimation at ABCC. Atomic Bomb Casualty Commission Technical Report 23–71, 1971
7. Kerr, G.D.: Organ dose estimates for the Japanese atomic bomb survivors. Health Phys 37:487–508, 1979
8. Milton, R.C. and Shohoji, T.: Tentative 1965 radiation dose estimation for atomic bomb survivors, Hiroshima and Nagasaki. ABCC Technical Report 1–68, 1968
9. Moloney, W.C. and Kastenbaum, M.A.: Leukemogenic effects of ionizing radiation on atomic bomb survivors in Hiroshima city. Science 121:308–309, 1954
10. Neel, J.V. and Schull, W.J.: *The effect of exposure to the atomic bombs on pregnancy termination in Hiroshima and Nagasaki*. Washington, National Academy of Sciences-National Research Council Publ. No. 461, 1956
11. Ritchie, R.H. and Hurst, G.S.: Penetration of weapons radiation: Application to the Hiroshima and Nagasaki studies. Health Phys 1:390–404, 1959
12. Roesch, W.C. (Editor): *Reassessment of Atomic Bomb Radiation Dosimetry in Hiroshima and Nagasaki: Final Report.* Hiroshima, Radiation Effects Research Foundation, 1987
13. Shimizu, Y., Kato, H., Schull, W.J., Preston, D.L., Fujita, S., and Pierce, D.A.: Life Span Study Report 11. Part 1. Comparison of the risk coefficients for site-specific cancer mortality based on the DS86 and T65D shielded kermas and organ doses. Radiation Effects Research Foundation Technical Report, 1987
14. UNSCEAR: *Biological Effects of Pre-natal Irradiation.* New York, United Nations, 1986

Discussion

C. MAYS: I'd like to make a comment. I'm very much impressed by the work that has been done by Drs. Otake and Schull with respect to the effects of radiation affecting human intelligence. To me this is certainly the most interesting development that has occurred in the understanding of radiobiology in the last 20 years. It also points out, I think, a defect in the glib use of the words "stochastic" and "non-stochastic" and the temptation to use these terms, which most of us have not understood, to force observations that are quite statistically valid into preconceived ideas that are not statistically or in any other way valid. The world "stochastic" has an appeal because it's a Greek word. It has two conditions on it: First, no threshold; second, the effect when induced is always equal in severity. Unfortunately, the term "non-stochastic" as defined by ICRP is actually "anti-stochastic", it is the mirror opposite of these two conditions and it is possible, in fact, to have 2 other type of non-stochastic effects. One, for example, in which there is a threshold and the effect does increase with increasing dose, in which category I would put effects on intelligence. I think that we should really look at the effects themselves without pre-existing constraints on what they're supposed to be, and try to really understand what they are. I'm quite persuaded that Dr. Schull is correct, that it is not necessary to kill cells in the brain in order to affect human intelligence but simply to disrupt the timing of their migration patterns into areas which will permit the neurons to function at optimal efficiency.

W. WOOLSON: You mentioned (Dr. Schull) the LD50 estimate of $2\frac{1}{2}$ Gy, and I have several questions about that. Was DS86 used or scaled dosimetry? Which dose are you referring to, is it free-field or marrow? And then, were you careful to exclude other effects like blast and thermal?

W. SCHULL: It's DS86. Those are marrow doses. We cannot obviously totally exclude other blast effects. What we have tried to do, like blast effects, is to look at the effect upon the LD50 of integrating deaths over different periods of time from the first day on. And, at least for deaths occurring anytime after August 6 until the end of 1945, the LD50 does not change very much. As you know, the effects are so intimately confounded here, that we will never really be able to pull out all of the things we have. The data are restricted to individuals who were exposed in wooden buildings.

B. SCOTT: (Dr. Schull) What model was used to estimate the LD50?

W. SCHULL: I'm beginning to regret having introduced that footnote, because to account for what we did is fairly lengthy but let us say this: At specified doses or distance, because we actually tried to estimate the LD50 using marrow dose and using distance, we have fitted both linear models as well as an exponential decay in mortality and over the range of interest to us, the two predict almost equally well. These are data that reflect individuals exposed from 800 m to 1400 m, so we're really on the steep part of the slope.

H. ROSSI: Dr. Schull's remarks make it plain that there is not much of a chance of getting some information on neutron RBE's. But if any attempt is made at all, I hope it will not be made with the assumption that the RBE is constant with dose because we have a great deal of information to indicate that this is not the case. Any evidence like this should be based, I believe, on a constant times dose to the minus $^{1}/_{2}$ power of neutron dose. It's much more likely to be more realistic in terms of what we know of radiobiology. And in fact, a reasonable number for that constant, I would say is 5, if the dose is expressed in Gray. That would be a reasonable first try.

T. BROOKS: I'd like to ask Dr. Awa, in the individuals where you had very high frequencies of aberrations, like 40 percent, was there any evidence of the same type of aberration reoccurring which would indicate clones, and if that was the case, how did you handle those?

A. AWA: In the case of clone formers, in other words, the same keryotypical abnormalities, we have found only 3 cases in a total of several hundred cases. And we regarded these cases in a different analysis, the majority of cases showed a rather sporadic nature, it is hard to find clones. So that in the case of clones, we first exclude the number of cells carrying the same abnormalities and also subtract these numbers from the denominator. This is the way that we tried to calculate.

E. RADFORD: I'd like to ask either Dr. Schull or Dr. Preston. No mention has been made so far about the age-specific cancer risks, and I would ask on the basis of the new 1985 follow-up what would be the ratio of the excess risk for children under the age of 10, for all cancers except leukemia or including leukemia, compared to the average values that were given in their presentations?

D. PRESTON: I don't have anything here that would let me address that in any detail. The patterns seen are very similar to what was has been seen in the most recent life-span study report. The relative risk is higher in the youngest group and there appears to be some decrease as these people get older in these large relative risks. They begin to look, as they get older, more like what it looks like in persons who were exposed at older ages.

E. RADFORD: In the published data that is already in the literature, it shows clearly that in equivalent attained ages, the risks are still substantially higher in the zero to ten group and I'm just curious as to how well this is holding past 1982, for example. Is it holding reasonably similar?

D. PRESTON: It's very similar to what it was in Report 10. The younger people are at slightly higher risk but the really large excess you saw in the youngest group is that their relative risk decrease markedly, now it's sort of flattening out.

E. RADFORD: Perhaps a factor of 2 then compared to the average?

D. PRESTON: I'm not going to say what the numbers are.

W. SCHULL: I found some numbers, I'm not sure these are exactly the ones that Dr. Radford is looking for, because as I tried to emphasize, we have not completed that part. Basically what we have looked at, is to simply take ages under 20 as opposed to over 20 as a first pass. If you do that and look at the ratio of the relative risk under 20 to that over 20, that ratio is larger under the DS86 than it is under the T65, but it is not very much larger: 5 percent.

N. WALD: Has there been any analysis of the cancer incidence in the population in which you scored aberrations and is there any difference between the incidence in the high aberration frequency individuals as compared to the low aberration frequency individuals in the same dose range, because we did see quite a range?

A. AWA: This kind of analysis is now in progress. So far we don't have data in terms of high aberration frequency cases. They may have higher cancer mortality but this kind of analysis is on the way. Maybe later I can answer this question, so far I don't have data with me.

D. KAUL: I have heard two comments so far, one that there is statistically significant difference between cities, and another from, I think, Dr. Schull, that there isn't statistical significant difference between cities. This may differ from effect to effect, but it clearly indicates that things are close. What I have not heard mentioned is any indication of the use of the dose uncertainties to further fuzz this comparison; and I would be interested to know whether any effort to take those uncertainties into account is being made, or if you are waiting for those uncertainties to be changed such that they include some kind of correlative information not presently in them. What is it you're waiting for?

D. PRESTON: The issue of dealing with the uncertainties is a very difficult one, as you know. We are making efforts, it does increase the variance, I think, and that is well known, and it does as you say, "fuzz" the comparison, but I think it's also a very difficult question. Efforts are underway to consider various aspects of the uncertainty. As far as the DS86 system, it provides uncertainties on about 20 percent of the survivors. We could use those to come up with uncertainties for others and they can be used to inflate the variance. They have not been in the results we presented today. But, I think as we move along in this, we have only just begun working with these doses, and there are other questions of bias, that I think are much more serious than even the variance inflation, that are not being addressed in the dosimetry evaluation and are probably not an appropriate part of the reassessment. But those effects are probably even greater because they affect not only the variance estimates but also the basic parameter as to the risk estimates themselves. I think this is a really important issue that is being looked at fairly seriously in recent months and I think a larger effort will be made there as time goes on.

D. KAUL: I bring the subject up because the subject of uncertainties in DS86 as it applies to individual survivors is not finished by any means. The issue of what kind of uncertainties to present and provide as part of the information that comes out of DS86 is still under discussion. The level of uncertainty that we've put into DS86 is really very simple at this time, but we have not considered such things as quantifying survivor-survivor correlations at a given city to try to understand how cohort migration might occur with these uncertainties or the question of city correlation to try to deal with these issues. All of that is under discussion and the architecture for providing that kind of information, which we have been goaded to do by Seymour Jablon, is still all in a formative stage; so people who have positions on these things, people that have desires in terms of being able to do their analysis and not have some kind of amorphous uncertainty that they have to deal with, ought to bring their desires to our attention as soon as they can. I have one more question and that is to do with the pregnant phantom. There is no pregnant phantom *per se* in DS86 right now. It is my understanding that the team at Oak Ridge Associated Universities has designed a first trimester phantom which we are trying to obtain to integrate into DS86. It looks to me, however, based on presentation of Dr. Otake, that there is a need for a second trimester phantom. Is this true? Do you need two pregnant phantoms as opposed to one?

W. SCHULL: If George Kerr is here he might want to enlarge on it, but George has asserted to us that for the second half of pregnancy there is a very close correspondence between the uterus dose and the absorbed dose of the fetus itself. For the first half of pregnancy, and one that is really critical to us, where the fetus is surrounded by much more amniotic fluid and therefore, the possibility of more absorption, we are not really sure how good the uterus is as a model. And I personally would very much like to see a model that looked at the fetus of an age that appears to be commensurate with where we see the major period of brain vulnerability.

D. KAUL: Which is?

W. SCHULL: 8 to 15 weeks.

D. KAUL: While that may be true about where the dose is being deposited or what it may be correlated with, there is a difference between the mass that is protecting the fetus or the uterus in the two phantoms so that needs to be taken into account.

J. WALKER: I'd like to direct this question to Dr. Schull. A lot of attention has been made on the direct effects of the radiation on the fetus. Now I was wondering if you could discuss the possibility of such things as microcephaly or mental retardation due to the radiation affecting the structures of the uterus from the mother being exposed.

W. SCHULL: My inclination is to refer you to some of the published work of Dr. Otake and I where we have tried to identify first what the concomitants might be that could color these risks. Clearly, most of them relate to the mother, the state of her health, the nutritional status at the time of the bombing, etc. While we cannot address any one of those issues as perfectly as we would like, nonetheless, we see no compelling evidence to believe that matronly mediated effects are the primary or even a large contributor to the dose response relationships that we see here. You'll find the arguments in the references set out much more completely in our publications and I would also get in a little advertising and suggest that you look at the new ICRP task group document which deals with the effects on the brain *in utero* exposure.

E. WEBSTER: I would like to address Dr. Otake's paper also, partly because I had a hard time getting data that I wanted out of the tables.

But there was one graphical slide relating to the rather critical 8 to 15 week group. It seemed to me that if you compare the lowest dose group, below 10 rads, with the next dose group, the percent incidence of severe mental retardation was about the same. And I wonder if that has any bearing on the fact that you did find a significant square law term and you could fit, in fact, a linear quadratic to that particular set of data?

W. SCHULL: Let me make this observation, I think it is true that under the DS86 at face value, there is not the same evidence for linearity that existed under the T65, at very low doses. I think that is also supported in part by the recognition that the Lee model fits a bit better than the linear one, but we can't reject either of them, nor can we discriminate between them statistically.

E. WEBSTER: May I follow up on that? If you limited the data to below 50 rads, would that make a difference? Because the curve looked very flat between 5 rads and 23 rads which were those bottom two data points on the block diagram.

W. SCHULL: We haven't looked at that, but I think you must bear in mind that in this sample there are only 30 severely retarded children and 9 of those occur in the non-exposed group. The moment that you ask of us to partition in more extensive ways, we have real problems, there aren't that many children there. I think if we could eventually evolve some sort of a metric which took all of the cerebral performance that we have (mental retardation, IQ and school performance) into account, then we might be able to look at something like that, and we could come up with an index of harm for the brain.

W. SINCLAIR: I want to ask Dale Preston a question: Yesterday the risk figure for non-leukemia was given as 0.7 per Sv. I wondered if he'd be willing to translate that himself into excess cases per million per Sv which seems to translate into a moderately high number. And if that's the case, since it's based on organ dose, and we've heard that DS86 and T65 don't differ very much in organ dose, where has the excess come from, and is it in the Reports 9, 10, and 11 that follow the BEIR Report data?

D. PRESTON: If you compare with Report 10 or 9, they are based on kerma, so there's going to be a shift when we move to organ dose, that's part of what you're seeing. I think that's the main thing, if you're comparing to earlier RERF published reports.

W. SINCLAIR: Well, that's about a factor of 2, but we're talking about more than that, aren't we? The first part of my question, would you translate the 0.7 per Sv?

D. PRESTON: I think that there has been very little change in the relative risk figures for non-leukemia between Report 10, and the current analysis and so if we work from this, if we look at all cancers except leukemia with T65D we will see a relative risk of about 0.27 using the methods used in Reports 9 and 10, and that's based on kerma. If we then translate that to organ dose, we see a shift to about 0.7 using T65D and George Kerr's transmission factors for the large intestine.

W. SINCLAIR: How much do you translate that into in excess cases per million?

D. PRESTON: I haven't done this, I'll try to do a calculation.

J. PUSKIN: I noticed in Dr. Otake's talk that the period of gestation relative to the bomb was defined by taking the difference from the date of birth. The question is: Do you have any ancillary information that could be used to pinpoint that better and is the sensitive period likely to be appreciably shorter than the 8 weeks that it currently looks like?

W. SCHULL: Our inability to date gestational ages better is a troublesome feature. We are dependent upon historical information, we do have information on the stated birth weights so we could at least use that to determine a term pregnancy or not. But, given the circumstances obtained at the time, there were, and will always be ambiguities in these data in that respect. Ideally, one would like to see a similar experiment, but if the Russians had been foresighted enough to do ultrasound on pregnant women so they could have dated the stage of pregnancy at which they were exposed in the Chernobyl accident, we might have a better notion of what's happening. It is obviously important to bear in mind that even if we could estimate gestational age precisely, what we're really interested in is developmental age. And we know that there is substantial variation between human fetuses of the same ostensible gestational age. Ideally we'd have a measure of development at the time that they were exposed, I don't see any way that we're going to get at that. In fact, I've often felt that perhaps some of this effect that we see at 16 to 25 weeks may really reflect the inaccuracies in our gestational age estimates.

Reports on Selected NCRP Activities

Warren K. Sinclair
Chairman

Scientific Committee 61— The Measurement of Radon and Radon Daughters in Air

Naomi H. Harley, *Chairman*
Dept. of Environmental Medicine
New York University Medical Center

The purpose of this committee was to prepare a comprehensive report on the measurement of radon and radon daughters. There are many ways to perform measurements of radon or daughters and in the hands of a competent scientist high quality data can be obtained. This report is not intended to be a cookbook describing step by step how a measurement is made but rather to describe the many available measurement techniques in detail and any advantages or disadvantages associated with particular protocols.

There are 13 chapters, 7 devoted to the measurement techniques and 6 describing background information. The background material includes the physical properties of radon and radon daughters, distribution of uranium, thorium, radium and radon in the environment, radon emanation and transport, behavior of the short-lived daughters in indoor environments, and formation of the unattached and attached short-lived daughters. The measurement sections deal with the determination of radon gas, individual radon daughters, potential alpha energy concentration (working level), unattached daughters and flux determination. For radon gas, there are 3 methods to obtain instantaneous or grab samples, 4 methods of continuous monitoring and 4 methods for obtaining time-averaged data. Twelve methods for measurement of individual radon daughters and 2 for thoron daughters are described. The measurement of potential alpha energy concentration (working level) has 6 techniques and 2 for personal monitoring. Four types of samplers for unattached daughters are available.

All of the radon or radon daughter measurement techniques described are in use in environmental or occupational situations. The only surprising finding was that essentially all measurements of the unattached fraction are probably somewhat in error. The unattached fraction is important because this small fraction of radioactivity deposits efficiently in the upper airways and delivers a disproportionately large fraction of the total alpha dose (30%). The measurements of unattached fraction have relied upon a knowledge of their diffusion coefficient. Because the particles grow from the moment of formation, the assumptions may have been in error.

Stress is placed on the quality of the measurements and upon quality control in general. One chapter is devoted to quality control in a measurement program. Without quality control little reliance can be placed upon the data.

The final chapter is a strategy chapter. There are 4 reasons for making radon or daughter measurements, evaluation of human exposure, identification of areas of high indoor levels, scientific studies and diagnostic measurements when carrying out remedial action. The different strategies for the different programs are discussed in detail. Radon versus radon daughter measurements are weighed, reporting of data, sampling location and methods in surveys for distributional studies. Radon is considered the preferred measurement over radon daughters because there is currently no standard atmosphere or intercomparison program for radon daughters as there is for radon measurements.

Scientific Committee 70—Quality Assurance in Diagnostic Radiology

Andrew K. Poznanski, *Chairman*
Children's Memorial Hospital
Chicago, Illinois

The Committee 70 has completed the writing of a publication on "Quality Assurance in Diagnostic Radiology." The purpose of this was to encourage establishment of practical comprehensive quality assurance programs in medical facilities, as well as to provide guidance to those engaged in developing and implementing such programs. The publication is intended primarily for the practicing diagnostic radiologist, other physicians, and medical physicists.

The outline of the publication is as follows:

0. Introduction
1. General Concepts of Quality Assurance and Quality Control
2. Quality Assurance Personnel Performance
3. Establishment of a Quality Control Program
4. Procedures, Objectives, and Policies
5. Photographic Quality Control
6. Conventional Radiography
7. Fluoroscopy and Cine
8. Mobile Systems
9. Tomography
10. Mammography

11. Dental Radiography
12. Special Procedures
13. Computed Tomography
14. Digital Imaging
15. Nuclear Medicine
16. Diagnostic Ultrasound
17. Magnetic Resonance Imaging
18. Video Systems
19. Computers
20. Miscellaneous
Appendix

Both the general concepts of QA and QC were discussed in the publication, including the definitions, approaches, and objectives, etc. There are also parts on cost-benefit considerations in education of various health care personnel. The booklet also deals with personnel performance as well as performance of equipment.

There is a chapter on various policy statements dealing with various aspects of the functioning of a radiology department. These are listed as Procedures, Objectives, Suggested Policies, and Comments. The Appendix includes an outline of the various tests that need to be done on various equipment, including reference as to where this is further discussed in the text, whether the test is essential or desirable, the test devices that must be used, suggested performance criteria, and minimum frequency that the tests should be performed.

Task Group 11 on Genetic Risk, Scientific Committee 57 on Internal Emitter Standards—Genetic Effects from Internally Deposited Radionuclides

Antone L. Brooks, *Chairman,*
Inhalation Toxicology Research Institute
Albuquerque, New Mexico

In this report, the data available to estimate the genetic effects from internally deposited radionuclides were evaluated. Since there is no information on the genetic effects of internally deposited radionuclides in man, our approach was to compare genetic change induced in a variety of experimental systems from both internally deposited radionuclides and external radiation exposure. From these comparisons, ratios of effect were established between the genetic change induced by external radiation and that observed for internally deposited radioactive materials. These ratios are similar to RBE and were used to predict relative effectiveness for the induction of genetic change by the different radiation types from internally deposited radioactive materials. The report was organized according to the characteristics of the radiation types and the radionuclides that produce these emissions.

For the low-energy beta emitters both carbon-14 and tritium were evaluated and effectiveness relative to low dose rate external exposure was taken to be 1.0 and 2.0, respectively. The high energy beta emitters were not found to produce unique genetic damage relative to external radiation and were assigned an effectiveness factor of 1.0. For internally deposited alpha emitting radionuclides it was determined that for induction of chromosome damage they were 15–20 times as effective as external x- or gamma rays. For the induction of transmitted mutations the effectiveness for alpha particles was 2.5–10 times as effective as external x- or gamma rays. It was also noted that specific-locus mutations induced by alpha irradiation appear to be qualitatively more severe than those induced by gamma rays. In light of the wide range of effectiveness observed for alpha particles for the different measures of genetic damage, it was recommended that a conservative approach be used and that an overall quality factor of 20 is appropriate for genetic damage produced by alpha irradiation relative to protracted low-LET radiation.

Taken as a whole, the data seem to suggest that there is little evidence for a unique genetic hazard from internally deposited radioactive material which would not be predicted by external exposure studies.

NCRP Publications

NCRP publications are distributed by the NCRP Publications' office. Information on prices and how to order may be obtained by directing an inquiry to:

NCRP Publications
7910 Woodmont Ave., Suite 1016
Bethesda, MD 20814

The currently available publications are listed below.

Proceedings of the Annual Meeting

No.	Title
1	*Perceptions of Risk,* Proceedings of the Fifteenth Annual Meeting, Held on March 14-15, 1979 (Including Taylor Lecture No. 3) (1980)
2	*Quantitative Risk in Standards Setting,* Proceedings of the Sixteenth Annual Meeting, Held on April 2-3, 1980 (Including Taylor Lecture No. 4) (1981)
3	*Critical Issues in Setting Radiation Dose Limits,* Proceedings of the Seventeenth Annual Meeting, Held on April 8-9, 1981 (Including Taylor Lecture No. 5) (1982)
4	*Radiation Protection and New Medical Diagnostic Procedures,* Proceedings of the Eighteenth Annual Meeting, Held on April 6-7, 1982 (Including Taylor Lecture No. 6) (1983)
5	*Environmental Radioactivity,* Proceedings of the Nineteenth Annual Meeting, Held on April 6-7, 1983 (Including Taylor Lecture No. 7) (1984)
6	*Some Issues Important in Developing Basic Radiation Protection Recommendations,* Proceedings of the Twentieth Annual Meeting, Held on April 4-5, 1984 (Including Taylor Lecture No. 8) (1985)
7	*Radioactive Waste,* Proceedings of the Twenty-First Annual Meeting, Held on April 3-4, 1985 (Including Taylor Lecture No. 9) (1986)

8 *Nonionizing Electromagnetic Radiations and Ultrasound,* Proceedings of the Twenty-second Annual Meeting, Held on April 2-3, 1986 (Including Taylor Lecture No. 10 [1988])

9 *New Dosimetry at Hiroshima and Nagasaki and Its Implications for Risk Estimates,* Proceedings of the Twenty-third Annual Meeting, Held on April 5-6, 1987 (Including Taylor Lecture No. 11) (1988)

Symposium Proceedings

The Control of Exposure of the Public to Ionizing Radiation in the Event of Accident or Attack, Proceedings of a Symposium held April 27–29, 1981 (1982)

Lauriston S. Taylor Lectures

No.	Title and Author
1	*The Squares of the Natural Numbers in Radiation Protection* by Herbert M. Parker (1977)
2	*Why be Quantitative About Radiation Risk Estimates?* by Sir Edward Pochin (1978)
3	*Radiation Protection—Concepts and Trade Offs* by Hymer L. Friedell (1979) [Available also in *Perceptions of Risk,* see above]
4	*From "Quantity of Radiation" and "Dose" to "Exposure" and "Absorbed Dose"—An Historical Review* by Harold O. Wyckoff (1980) [Available also in *Quantitative Risks in Standards Setting,* see above]
5	*How Well Can We Assess Genetic Risk? Not Very* by James F. Crow (1981) [Available also in *Critical Issues in Setting Radiation Dose Limits,* see above]
6	*Ethics, Trade-offs and Medical Radiation* by Eugene L. Saenger (1982) [Available also in *Radiation Protection and New Medical Diagnostic Approaches,* see above]
7	*The Human Environment-Past, Present and Future* by Merril Eisenbud (1983) [Available also in *Environmental Radioactivity,* see above]

8 *Limitation and Assessment in Radiation Protection* by Harald H. Rossi (1984) [Available also in *Some Issues Important in Developing Basic Radiation Protection Recommendations*, see above]

9 *Truth (and Beauty) in Radiation Measurement* by John H. Harley (1985) [Available also in *Radioactive Waste*, see above]

10 *Nonionizing Radiation Bioeffects: Cellular Properties and Interactions* by Herman P. Schwan (1987) [Available also in *Nonionizing Electromagnetic Radiations and Ultrasound*, see above]

11 *How to be Quantitative about Radiation Risk Estimates* by Semour Jablon (1987) [Available also in *New Dosimetry at Hiroshima and Nagasaki and Its Implications for Risk Estimates*, (see above)]

12 *How Safe is Safe Enough?* by Bo Lindell (1988)

NCRP Commentaries

Commentary No.	Title
1	*Krypton-85 in the Atmosphere—With Specific Reference to the Public Health Significance of the Proposed Controlled Release at Three Mile Island* (1980)
2	*Preliminary Evaluation of Criteria for the Disposal of Transuranic Contaminated Waste* (1982)
3	*Screening Techniques for Determining Compliance with Environmental Standards* (1986)
4	*Guidelines for the Release of Waste Water from Nuclear Facilities with Special Reference to the Public Health Significance of the Proposed Release of Treated Waste Waters at Three Mile Island* (1987)

NCRP Reports

No.	Title
8	*Control and Removal of Radioactive Contamination in Laboratories* (1951)
16	*Radioactive Waste Disposal in the Ocean* (1954)
22	*Maximum Permissible Body Burdens and Maximum Permissible Concentrations of Radionuclides in Air and in*

Water for Occupational Exposure (1959) [Includes Addendum 1 issued in August 1963]

23 *Measurement of Neutron Flux and Spectra for Physical and Biological Applications* (1960)

25 *Measurement of Absorbed Dose of Neutrons and Mixtures of Neutrons and Gamma Rays* (1961)

27 *Stopping Powers for Use with Cavity Chambers* (1961)

30 *Safe Handling of Radioactive Materials* (1964)

32 *Radiation-Protection in Educational Institutions* (1966)

33 *Medical X-Ray and Gamma-Ray Protection for Energies Up to 10 MeV—Equipment Design and Use* (1968)

35 *Dental X-Ray Protection* (1970)

36 *Radiation Protection in Veterinary Medicine* (1970)

37 *Precautions in the Management of Patients Who Have Received Therapeutic Amounts of Radionuclides* (1970)

38 *Protection Against Neutron Radiation* (1971)

40 *Protection Against Radiation from Brachytherapy Sources* (1972)

41 *Specification of Gamma-Ray Brachytherapy Sources* (1974)

42 *Radiological Factors Affecting Decision-Making in a Nuclear Attack* (1974)

44 *Krypton-85 in the Atmosphere—Accumulation, Biological Significance, and Control Technology* (1975)

46 *Alpha-Emitting Particles in Lungs* (1975)

47 *Tritium Measurement Techniques* (1976)

48 *Radiation Protection for Medical and Allied Health Personnel* (1976)

49 *Structural Shielding Design and Evaluation for Medical Use of X Rays and Gamma Rays of Energies Up to 10 MeV* (1976)

50 *Environmental Radiation Measurements* (1976)

51 *Radiation Protection Design Guidelines for 0.1-100 MeV Particle Accelerator Facilities* (1977)

52 *Cesium-137 from the Environment to Man: Metabolism and Dose* (1977)

53 *Review of NCRP Radiation Dose Limit for Embryo and Fetus in Occupationally Exposed Women* (1977)

54 *Medical Radiation Exposure of Pregnant and Potentially Pregnant Women* (1977)

55 *Protection of the Thyroid Gland in the Event of Releases of Radioiodine* (1977)

57 *Instrumentation and Monitoring Methods for Radiation Protection* (1978)
58 *A Handbook of Radioactivity Measurements Procedures,* 2nd ed. (1985)
59 *Operational Radiation Safety Program* (1978)
60 *Physical, Chemical, and Biological Properties of Radiocerium Relevant to Radiation Protection Guidelines* (1978)
61 *Radiation Safety Training Criteria for Industrial Radiography* (1978)
62 *Tritium in the Environment* (1979)
63 *Tritium and Other Radionuclide Labeled Organic Compounds Incorporated in Genetic Material* (1979)
64 *Influence of Dose and Its Distribution in Time on Dose-Response Relationships for Low-LET Radiations* (1980)
65 *Management of Persons Accidentally Contaminated with Radionuclies* (1980)
66 *Mammography* (1980)
67 *Radiofrequency Electromagnetic Fields—Properties, Quantities and Units, Biophysical Interaction, and Measurements* (1981)
68 *Radiation Protection in Pediatric Radiology* (1981)
69 *Dosimetry of X-Ray and Gamma-Ray Beams for Radiation Therapy in the Energy Range 10 keV to 50 MeV* (1981)
70 *Nuclear Medicine-Factors Influencing the Choice and Use of Radionuclides in Diagnosis and Therapy* (1982)
71 *Operational Radiation Safety—Training* (1983)
72 *Radiation Protection and Measurement for Low Voltage Neutron Generators* (1983)
73 *Protection in Nuclear Medicine and Ultrasound Diagnostic Procedures in Children* (1983)
74 *Biological Effects of Ultrasound: Mechanisms and Clinical Implications* (1983)
75 *Iodine-129: Evaluation of Releases from Nuclear Power Generation* (1983)
76 *Radiological Assessment: Predicting the Transport, Bioaccumulation, and Uptake by Man of Radionuclides Released to the Environment* (1984)
77 *Exposures from the Uranium Series with Emphasis on Radon and its Daughters* (1984)

78 *Evaluation of Occupational and Environmental Exposures to Radon and Radon Daughters in the United States* (1984)

79 *Neutron Contamination from Medical Electron Accelerators* (1984)

80 *Induction of Thyroid Cancer by Ionizing Radiation* (1985)

81 *Carbon-14 in the Environment* (1985)

82 *SI Units in Radiation Protection and Measurements* (1985)

83 *The Experimental Basis for Absorbed-Dose Calculations in Medical uses of Radionuclides* (1985)

84 *General Concepts for the Dosimetry of Internally Deposited Radionuclides* (1985)

85 *Mammography—A User's Guide* (1986)

86 *Biological Effects and Exposure Criteria for Radiofrequency Electromagnetic Fields* (1986)

87 *Use of Bioassay Procedures for Assessment of Internal Radionuclide Deposition* (1986)

88 *Radiation Alarms and Access-Control Systems* (1987)

89 *Genetic Effects of Internally Deposited Radionuclides* (1987)

90 *Neptunium: Radiation Protection Guidelines* (1987)

91 *Recommendations on Limits for Exposure to Ionizing Radiation* (1987)

92 *Public Radiation Exposure from Nuclear Power Generation in the United States* (1987)

93 *Ionizing Radiation Exposure of the Population of the United States* (1987)

94 *Exposure of the Population in the United States and Canada from Natural Background Radiation* (1987)

95 *Radiation Exposure of the U.S. Population from Consumer Products and Miscellaneous Sources* (1987)

96 *Comparative Carcinogenesis of Ionizing Radiation and Chemicals* (1988)

97 *Measurement of Radon and Radon Daughters in Air* (1988)

98 *Guidance on Radiation Received in Space Activities* (1988)

99 *Quality Assurance For Diagnostic Imaging Equipment* (1988)

100 *Exposure of the U.S. Population from Diagnostic Medical Radiation* (1988)

Binders for NCRP Reports are available. Two sizes make it possible to collect into small binders the "old series" of reports (NCRP Reports Nos. 8-30) and into large binders the more recent publications (NCRP Reports Nos. 32-96). Each binder will accommodate from five to seven reports. The binders carry the identification "NCRP Reports" and come with label holders which permit the user to attach labels showing the reports contained in each binder.

The following bound sets of NCRP Reports are also available:

Volume I. NCRP Reports Nos. 8, 16,22
Volume II. NCRP Reports Nos. 23, 25, 27, 30
Volume III. NCRP Reports Nos. 32, 33, 35, 36, 37
Volume IV. NCRP Reports Nos. 38, 40, 41
Volume V. NCRP Reports Nos. 42, 44, 46
Volume VI. NCRP Reports Nos. 47, 48, 49, 50, 51
Volume VII. NCRP Reports Nos. 52, 53, 54, 55, 57
Volume VIII. NCRP Reports No. 58
Volume IX. NCRP Reports Nos. 59, 60, 61, 62, 63
Volume X. NCRP Reports Nos. 64, 65, 66, 67
Volume XI. NCRP Reports Nos. 68, 69, 70, 71, 72
Volume XII. NCRP Reports Nos. 73, 74, 75, 76
Volume XIII. NCRP Reports Nos. 77, 78, 79, 80
Volume XIV. NCRP Reports Nos. 81, 82, 83, 84, 85
Volume XV. NCRP Reports Nos. 86, 87, 88, 89
Volume XVI. NCRP Reports Nos. 90, 91, 92, 93

(Titles of the individual reports contained in each volume are given above.)

The following NCRP Reports are now superseded and/or out of print:

No.	Title
1	*X-Ray Protection* (1931). [Superseded by NCRP Report No. 3]
2	*Radium Protection* (1934). [Superseded by NCRP Report No. 4]
3	*X-Ray Protection* (1936). [Superseded by NCRP Report No. 6]
4	*Radium Protection* (1938). [Superseded by NCRP Report No. 13]
5	*Safe Handling of Radioactive Luminous Compounds* (1941). [Out of Print]

6 *Medical X-Ray Protection Up to Two Million Volts* (1949). [Superseded by NCRP Report No. 18]

7 *Safe Handling of Radioactive Isotopes* (1949). [Superseded by NCRP Report No. 30]

9 *Recommendations for Waste Disposal of Phosphorus-32 and Iodine-131 for Medical Users* (1951). [Out of Print]

10 *Radiological Monitoring Methods and Instruments* (1952). [Superseded by NCRP Report No. 57]

11 *Maximum Permissible Amounts of Radioisotopes in the Human Body and Maximum Permissible Concentrations in Air and Water* (1953). [Superseded by NCRP Report No. 22]

12 *Recommendations for the Disposal of Carbon-14 Wastes* (1953). [Superseded by NCRP Report No. 81]

13 *Protection Against Radiations from Radium, Cobalt-60 and Cesium-137* (1954). [Superseded by NCRP Report No. 24]

14 *Protection Against Betatron-Synchrotron Radiations Up to 100 Million Electron Volts* (1954). [Superseded by NCRP Report No. 51]

15 *Safe Handling of Cadavers Containing Radioactive Isotopes* (1953). [Superseded by NCRP Report No. 21]

17 *Permissible Dose from External Sources of Ionizing Radiation* (1954) including *Maximum Permissible Exposure to Man, Addendum to National Bureau of Standards Handbook 59* (1958). [Superseded by NCRP Repopt No. 39]

18 *X-Ray Protection* (1955). [Superseded by NCRP Report No. 26]

19 *Regulation of Radiation Exposure by Legislative Means* (1955). [Out of Print]

20 *Protection Against Neutron Radiation Up to 30 Million Electron Volts* (1957). [Superseded by NCRP Report No. 38]

21 *Safe Handling of Bodies Containing Radioactive Isotopes* (1958). [Superseded by NCRP Report No. 37]

24 *Protection Against Radiations from Sealed Gamma Sources* (1960). [Superseded by NCRP Report Nos. 33, 34, and 40]

26 *Medical X-Ray Protection Up to Three Million Volts* (1961). [Superseded by NCRP Report Nos. 33, 34, 35, and 36]

28 *A Manual of Radioactivity Procedures* (1961). [Superseded by NCRP Report No. 58]

29 *Exposure to Radiation in an Emergency* (1962). [Superseded by NCRP Report No. 42]

31 *Shielding for High Energy Electron Accelerator Installations* (1964). [Superseded by NCRP Report No. 51]

34 *Medical X-Ray and Gamma-Ray Protection for Energies Up to 10 MeV—Structural Shielding Design and Evaluation* (1970). [Superseded by NCRP Report No. 49]

39 *Basic Radiation Protection Criteria* (1971). [Superseded by NCRP Report No. 91]

43 *Review of the Current State of Radiation Protection Philosophy* (1975). [Superseded by NCRP Report No. 91]

45 *Natural Background Radiation in the United States* (1975). [Superseded by NCRP Report No. 94]

56 *Radiation Exposure from Consumer Products and Miscellaneous Sources* (1977). [Superseded by NCRP Report No. 95]

58 *A Handbook of Radioactivity Measurement Procedures.* [Superseded by NCRP Report No. 58 2nd ed.]

Other Documents

The following documents of the NCRP were published outside of the NCRP Reports and Commentaries series:

"Blood Counts, Statement of the National Committee on Radiation Protection," Radiology 63, 428 (1954)

"Statements on Maximum Permissible Dose from Television Receivers and Maximum Permissible Dose to the Skin of the Whole Body," Am. J. Roentgenol., Radium Ther. and Nucl. Med. 84, 152 (1960) and Radiology 75, 102 (1960)

Dose Effect Modifying Factors In Radiation Protection, Report of Subcommittee M-4 (Relative Biological) Effectiveness of the National Council on Radiation Protection and Measurements, Report BNL 50073 (T-471) Brookhaven National Laboratory (1967). Available from National Technical Information Service, Springfield, Virginia

X-Ray Protection Standards for Home Television Receivers, Interim Statement of the National Council on Radiation Protection and Measurements (National Council on Radiation Protection and Measurements, Washington, 1968)

Specification of Units of Natural Uranium and Natural Thorium (National Council on Radiation Protection and Measurements, Washington, 1973)

NCRP Statement on Dose Limit for Neutrons (National Council on Radiation Protection and Measurements, Washington, 1980)

Control of Air Emissions of Radionuclides (National Council on Radiation Protection and Measurements, Bethesda, Maryland, 1984)

Copies of the statements published in journals may be consulted in libraries. A limited number of copies of the remaining documents listed above are available for distribution by NCRP Publications.